tredition®

tredition was established in 2006 by Sandra Latusseck and Soenke Schulz. Based in Hamburg, Germany, tredition offers publishing solutions to authors and publishing houses, combined with worldwide distribution of printed and digital book content. tredition is uniquely positioned to enable authors and publishing houses to create books on their own terms and without conventional manufacturing risks.

For more information please visit: www.tredition.com

TREDITION CLASSICS

This book is part of the TREDITION CLASSICS series. The creators of this series are united by passion for literature and driven by the intention of making all public domain books available in printed format again - worldwide. Most TREDITION CLASSICS titles have been out of print and off the bookstore shelves for decades. At tredition we believe that a great book never goes out of style and that its value is eternal. Several mostly non-profit literature projects provide content to tredition. To support their good work, tredition donates a portion of the proceeds from each sold copy. As a reader of a TREDITION CLASSICS book, you support our mission to save many of the amazing works of world literature from oblivion. See all available books at www.tredition.com.

 Project Gutenberg

The content for this book has been graciously provided by Project Gutenberg. Project Gutenberg is a non-profit organization founded by Michael Hart in 1971 at the University of Illinois. The mission of Project Gutenberg is simple: To encourage the creation and distribution of eBooks. Project Gutenberg is the first and largest collection of public domain eBooks.

Station Life in New Zealand

Lady (Mary Anne) Barker

Imprint

This book is part of TREDITION CLASSICS

Author: Lady (Mary Anne) Barker
Cover design: Buchgut, Berlin – Germany

Publisher: tredition GmbH, Hamburg - Germany
ISBN: 978-3-8424-6107-9

www.tredition.com
www.tredition.de

Copyright:
The content of this book is sourced from the public domain.

The intention of the TREDITION CLASSICS series is to make world literature in the public domain available in printed format. Literary enthusiasts and organizations, such as Project Gutenberg, worldwide have scanned and digitally edited the original texts. tredition has subsequently formatted and redesigned the content into a modern reading layout. Therefore, we cannot guarantee the exact reproduction of the original format of a particular historic edition. Please also note that no modifications have been made to the spelling, therefore it may differ from the orthography used today.

STATION LIFE IN NEW ZEALAND

By Lady Barker.

1883

Contents

Preface.

Letter I. Two months at sea — Melbourne

Letter II. Sight-seeing in Melbourne

Letter III. On to New Zealand

Letter IV. First introduction to "Station life"

Letter V. A pastoral letter

Letter VI. Society — houses and servants

Letter VII. A young colonist — the town and its neighbourhood

Letter VIII. Pleasant days at Ilam

Letter IX. Death in our new home — New Zealand children

Letter X. Our station home

Letter XI. Housekeeping, and other matters

Letter XII. My first expedition

Letter XIII. Bachelor hospitality — a gale on shore

Letter XIV. A Christmas picnic, and other doings

Letter XV. Everyday station life

Letter XVI. A sailing excursion on Lake Coleridge

Letter XVII. My first and last experience of "camping out"

Letter XVIII. A journey "down south"

Letter XIX. A Christening gathering — the fate of Dick

Letter XX. the New Zealand snowstorm of 1867

Letter XXI.   Wild cattle hunting in the Kowai Bush

Letter XXII.   The exceeding joy of "burning"

Letter XXIII.   Concerning a great flood

Letter XXIV.   My only fall from horseback

Letter XXV.   How We lost our horses and had to walk home

Preface.

These letters, their writer is aware, justly incur the reproach of egotism and triviality; at the same time she did not see how this was to be avoided, without lessening their value as the exact account of a lady's experience of the brighter and less practical side of colonization. They are published as no guide or handbook for "the intending emigrant;" that person has already a literature to himself, and will scarcely find here so much as a single statistic. They simply record the expeditions, adventures, and emergencies diversifying the daily life of the wife of a New Zealand sheep-farmer; and, as each was written while the novelty and excitement of the scenes it describes were fresh upon her, they may succeed in giving here in England an adequate impression of the delight and freedom of an existence so far removed from our own highly-wrought civilization: not failing in this, the writer will gladly bear the burden of any critical rebuke the letters deserve. One thing she hopes will plainly appear, — that, however hard it was to part, by the width of the whole earth, from dear friends and spots scarcely less dear, yet she soon found in that new country new friends and a new home; costing her in their turn almost as many parting regrets as the old.

F. N. B.

Letter I: Two months at sea — Melbourne.

Port Phillip Hotel, Melbourne. September 22d, 1865. .... Now I must give you an account of our voyage: it has been a very quick one for the immense distance traversed, sometimes under canvas, but generally steaming. We saw no land between the Lizard and Cape Otway light — that is, for fifty-seven days: and oh, the monotony of that time! — the monotony of it! Our decks were so crowded that we divided our walking hours, in order that each set of passengers might have space to move about; for if every one had taken it into their heads to exercise themselves at the same time, we could hardly have exceeded the fisherman's definition of a walk, "two steps and overboard." I am ashamed to say I was more or less ill all the way, but, fortunately, F— — was not, and I rejoiced at this from the most selfish motives, as he was able to take care of me. I find that sea-sickness develops the worst part of one's character with startling rapidity, and, as far as I am concerned, I look back with self-abasement upon my callous indifference to the sufferings of others, and apathetic absorption in my individual misery.

Until we had fairly embarked, the well-meaning but ignorant among our friends constantly assured us, with an air of conviction as to the truth and wisdom of their words, that we were going at the very best season of the year; but as soon as we could gather the opinions of those in authority on board, it gradually leaked out that we really had fallen upon quite a wrong time for such a voyage, for we very soon found ourselves in the tropics during their hottest month (early in August), and after having been nearly roasted for three weeks, we plunged abruptly into mid-winter, or at all events very early spring, off the Cape of Good Hope, and went through a season of bitterly cold weather, with three heavy gales. I pitied the poor sailors from the bottom of my heart, at their work all night on decks slippery with ice, and pulling at ropes so frozen that it was almost impossible to bend them; but, thank God, there were no casualties among the men. The last gale was the most severe; they said it was the tail of a cyclone. One is apt on land to regard such phrases as the "shriek of the storm," or "the roar of the waves," as poetical hyperboles; whereas they are very literal and expressive renderings of the sounds of horror incessant throughout a gale at sea. Our cabin, though very nice and comfortable in other respects, possessed an extraordinary attraction for any stray wave which might be wandering about the saloon: once or twice I have been in the cuddy when a sea found its way down the companion, and I have watched with horrible anxiety a ton or so of water hesitating which cabin it should enter and deluge, and it always seemed to choose ours. All these miseries appear now, after even a few days of the blessed land, to belong to a distant past; but I feel inclined to lay my pen down and have a hearty laugh at the recollection of one cold night, when a heavy "thud" burst open our cabin door, and washed out all the stray parcels, boots, etc., from the corners in which the rolling of the ship had previously bestowed them. I was high and dry in the top berth, but poor F— — in the lower recess was awakened by the douche, and no words of mine can convey to you the utter absurdity of his appearance, as he nimbly mounted on the top of a chest of drawers close by, and crouched there, wet and shivering, handing me up a most miscellaneous assortment of goods to take care of in my little dry nest.

Some of our fellow-passengers were very good-natured, and devoted themselves to cheering and enlivening us by getting up concerts, little burlesques and other amusements; and very grateful we were for their efforts: they say that "anything is fun in the country," but on board ship a little wit goes a very long way indeed, for all are only too ready and anxious to be amused. The whole dramatic strength of the company was called into force for the performance of "The Rivals," which was given a week or so before the end of the voyage. It went off wonderfully well; but I confess I enjoyed the preparations more than the play itself: the ingenuity displayed was very amusing at the time. You on shore cannot imagine how difficult it was to find a snuff-box for "Sir Anthony Absolute," or with what joy and admiration we welcomed a clever substitute for it in the shape of a match-box covered with the lead out of a tea-chest most ingeniously modelled into an embossed wreath round the lid, with a bunch of leaves and buds in the centre, the whole being brightly burnished: at the performance the effect of this little "property" was really excellent. Then, at the last moment, poor "Bob Acres" had to give in, and acknowledge that he could not speak for coughing; he had been suffering from bronchitis for some days past, but had gallantly striven to make himself heard at rehearsals; so on the day of the play F— — had the part forced on him. There was no time to learn his "words," so he wrote out all of them in large letters on slips of paper and fastened them on the beams. This device was invisible to the audience, but he was obliged to go through his scenes with his head as high up as if he had on a martingale; however, we were all so indulgent that at any little *contretemps*, such as one of the actresses forgetting her part or being seized by stage-fright, the applause was much greater than when things went smoothly.

I can hardly believe that it is only two days since we steamed into Hobson's Bay, on a lovely bright spring morning. At dinner, the evening before, our dear old captain had said that we should see the revolving light on the nearest headland about eight o'clock that evening, and so we did. You will not think me childish, if I acknowledge that my eyes were so full of tears I could hardly see it after the first glimpse; it is impossible to express in a letter all the joy and thankfulness of such a moment. Feelings like these are forgotten only too quickly in the jar and bustle of daily life, and we are always ready to take as a matter of course those mercies which are new every morning; but when I realized that all the tosses and tumbles of so many weary days and nights were over, and that at last we had reached the haven where we would be, my first thought was one of deep gratitude. It was easy to see that it was a good moment with everyone; squabbles were made up with surprising quickness; shy people grew suddenly sociable; some who had comfortable homes to go to on landing gave kind and welcome invitations to others, who felt themselves sadly strange in a new country; and it was with really a lingering feeling of regret that we all separated at last, though a very short time before we should have thought it quite impossible to be anything but delighted to leave the ship.

We have not seen much of Melbourne yet, as there has been a great deal to do in looking after the luggage, and at first one is capable of nothing but a delightful idleness. The keenest enjoyment is a fresh-water bath, and next to that is the new and agreeable luxury of the ample space for dressing; and then it is so pleasant to suffer no anxiety as to the brushes and combs tumbling about. I should think that even the vainest woman in the world would find her toilet and its duties a daily trouble and a sorrow at sea, on account of the unsteadiness of all things. The next delight is standing at the window, and seeing horses, and trees, and dogs—in fact, all the "treasures of the land;" as for flowers—beautiful as they are at all times—you cannot learn to appreciate them enough until you have been deprived of them for two months.

You know that I have travelled a good deal in various parts of the world, but I have never seen anything at all like Melbourne. In other countries, it is generally the antiquity of the cities, and their historical reminiscences, which appeal to the imagination; but *here*, the interest is as great from exactly the opposite cause. It is most wonderful to walk through a splendid town, with magnificent public buildings, churches, shops, clubs, theatres, with the streets well paved and lighted, and to think that less than forty years ago it was a desolate swamp without even a hut upon it. How little an English country town progresses in forty years, and here is a splendid city created in that time! I have no hesitation in saying, that any fashionable novelty which comes out in either London or Paris finds its way to Melbourne by the next steamer; for instance, I broke my parasol on board ship, and the first thing I did on landing was to go to one of the best shops in Collins Street to replace it. On learning what I wanted, the shopman showed me some of those new parasols which had just come out in London before I sailed, and which I had vainly tried to procure in S— —, only four hours from London.

The only public place we have yet visited is the Acclimatization Garden; which is very beautifully laid out, and full of aviaries, though it looks strange to see common English birds treated as distinguished visitors and sumptuously lodged and cared for. Naturally, the Australian ones interest me most, and they are certainly prettier than yours at home, though they do not sing. I have been already to a shop where they sell skins of birds, and have half ruined myself in purchases for hats. You are to have a "diamond sparrow," a dear little fellow with reddish brown plumage, and white spots over its body (in this respect a miniature copy of the Argus pheasant I brought from India), and a triangular patch of bright yellow under its throat. I saw some of them alive in a cage in the market with many other kinds of small birds, and several pairs of those pretty grass or zebra paroquets, which are called here by the very inharmonious name of "budgerighars." I admired the blue wren so much—a tiny *birdeen* with tail and body of dust-coloured feathers, and head and throat of a most lovely turquoise blue; it has also a little wattle of these blue feathers standing straight out on each side of its head, which gives it a very pert appearance. Then there is the emu-wren, all sad-coloured, but quaint, with the tail-feathers sticking up on end, and exactly like those of an emu; on the very smallest scale, even to the peculiarity of two feathers growing out of the same little quill. I was much amused by the varieties of cockatoos, parrots, and lories of every kind and colour, shrieking and jabbering in the part of the market devoted to them; but I am told that I have seen very few of the varieties of birds, as it is early in the spring, and the young ones have not yet been brought in: they appear to sell as fast as they can be procured. But before I end my letter I must tell you about the cockatoo belonging to this hotel. It is a famous bird in its way, having had its portrait taken several times, descriptions written for newspapers of its talents, and its owner boasts of enormous sums offered and refused for it. Knowing my fondness for pets, F— — took me downstairs to see it very soon after our arrival. I thought it hideous: it belongs to a kind not very well known in England, of a dirtyish white colour, a very ugly-shaped head and bill, and large bluish rings round the eyes; the beak is huge and curved. If it knew of this last objection on my part, it would probably answer, like the wolf in Red Riding Hood's story, "the better to talk with, my dear"—for it is a weird and knowing bird. At first it flatly refused to show off any of its accomplishments, but one of the hotel servants good-naturedly came forward, and Cocky condescended to go through his performances. I cannot possibly-tell you of all its antics: it pretended to have a violent toothache, and nursed its beak in its claw, rocking itself backwards and forwards as if in the greatest agony, and in answer to all the remedies which were proposed, croaking out, "Oh, it ain't a bit of good," and finally sidling up, to the edge of its perch, and saying in hoarse but confidential whisper, "Give us a drop of whisky, *do*." Its voice was extraordinarily distinct, and when it sang several snatches of songs the words were capitally given, with the most absurdly comic intonation, all the *roulades* being executed in perfect tune. I liked its sewing performance so much—to see it hold a little piece of stuff underneath the claw which rested on the perch, and pretend to sew with the other, getting into difficulties with its thread, and finally setting up a loud song in praise of sewing-machines just as if it were an advertisement.

By the next time I write I shall have seen more of Melbourne; there will, however, be no time for another letter by this mail; but I will leave one to be posted after we sail for New Zealand.

Letter II: Sight-seeing in Melbourne.

Melbourne, October 1st, 1865. I have left my letter to the last moment before starting for Lyttleton; everything is re-packed and ready, and we sail to-morrow morning in the *Albion*. She is a mail-steamer—very small after our large vessel, but she looks clean and tidy; at all events, we hope to be only on board her for ten days. In England one fancies that New Zealand is quite close to Australia, so I was rather disgusted to find we had another thousand miles of steaming to do before we could reach our new home; and one of the many Job's comforters who are scattered up and down the world assures me that the navigation is the most dangerous and difficult of the whole voyage.

We have seen a good deal of Melbourne this week; and not only of the town, for we have had many drives in the exceedingly pretty suburbs, owing to the kindness of the D— —s, who have been most hospitable and made our visit here delightful. We drove out to their house at Toorak three or four times; and spent a long afternoon with them; and there I began to make acquaintance with the Antipodean trees and flowers. I hope you will not think it a very sweeping assertion if I say that all the leaves look as if they were made of leather, but it really is so; the hot winds appear to parch up everything, at all events, round Melbourne, till the greatest charm of foliage is more or less lost; the flowers also look withered and burnt up, as yours do at the end of a long, dry summer, only they assume this appearance after the first hot wind in spring. The suburb called Heidelberg is the prettiest, to my taste—an undulating country with vineyards, and a park-like appearance which, is very charming. All round Melbourne there are nice, comfortable, English-looking villas. At one of these we called to return a visit and found a very handsome house, luxuriously furnished, with beautiful garden and grounds. One afternoon we went by rail to St. Kilda's, a flourishing bathing-place on the sea-coast, about six miles from Melbourne. Everywhere building is going on with great rapidity, and you do not see any poor people in the streets. If I wanted to be critical and find fault, I might object to the deep gutters on each side of the road; after a shower of rain they are raging torrents for a short time, through which you are obliged to splash without regard to the muddy consequences; and even when they are dry, they entail sudden and prodigious jolts. There are plenty of Hansoms and all sorts of other conveyances, but I gave F— — no peace until he took me for a drive in a vehicle which was quite new to me—a sort of light car with a canopy and curtains, holding four, two on each seat, *dos-a-dos*, and called a "jingle,"—of American parentage, I fancy. One drive in this carriage was quite enough, however, and I contented myself with Hansoms afterwards; but walking is really more enjoyable than anything else, after having been so long cooped up on board ship.

We admired the fine statue, at the top of Collins Street, to the memory of the two most famous of Australian explorers, Burke and Wills, and made many visits to the Museum, and the glorious Free Library; we also went all over the Houses of Legislature—very new and grand. But you must not despise me if I confess to having enjoyed the shops exceedingly: it was so unlike a jeweller's shop in England to see on the counter gold in its raw state, in nuggets and dust and flakes; in this stage of its existence it certainly deserves its name of "filthy lucre," for it is often only half washed. There were quantities of emus' eggs in the silversmiths' shops, mounted in every conceivable way as cups and vases, and even as work-boxes: some designs consisted of three or five eggs grouped together as a centre-piece. I cannot honestly say I admired any of them; they were generally too elaborate, comprising often a native (spear in hand), a kangaroo, palms, ferns, cockatoos, and sometimes an emu or two in addition, as a pedestal—all this in frosted silver or gold. I was given a pair of these eggs before leaving England: they were mounted in London as little flower-vases in a setting consisting only of a few bulrushes and leaves, yet far better than any of these florid designs; but he emu-eggs are very popular in Sydney or Melbourne, and I am told sell rapidly to people going home, who take them as a memento of their Australian life, and probably think that the greater the number of reminiscences suggested by the ornament the more satisfactory it is as a purchase.

I must finish my letter by a description of a dinner-party which about a dozen of our fellow-passengers joined with us in giving our dear old captain before we all separated. Whilst we were on board, it very often happened that the food was not very choice or good: at all events we used sometimes to grumble at it, and we generally wound up our lamentations by agreeing that when we reached Melbourne we would have a good dinner together. Looking back on it, I must say I think we were all rather greedy, but we tried to give a better colouring to our gourmandism by inviting the captain, who was universally popular, and by making it as elegant and pretty a repast as possible. Three or four of the gentlemen formed themselves into a committee, and they must really have worked very hard; at all events they collected everything rare and strange in the way of fish, flesh, and fowl peculiar to Australia, the arrangement of the table was charming, and the delicacies were all cooked and served to perfection. The ladies' tastes were considered in the profusion of flowers, and we each found an exquisite bouquet by our plate. I cannot possibly give you a minute account of the whole menu; in fact, as it is, I feel rather like Froissart, who, after chronicling a long list of sumptuous dishes, is not ashamed to confess, "Of all which good things I, the chronicler of this narration, did partake!" The soups comprised kangaroo-tail—a clear soup not unlike ox-tail, but with a flavour of game. I wish I could recollect the names of the fish: the fresh-water ones came a long distance by rail from the river Murray, but were excellent nevertheless. The last thing which I can remember tasting (for one really could do little else) was a most exquisite morsel of pigeon—more like a quail than anything else in flavour. I am not a judge of wine, as you may imagine, therefore it is no unkindness to the owners of the beautiful vineyards which we saw the other day, to say that I do not like the Australian wines. Some of the gentlemen pronounced them to be excellent, especially the equivalent to Sauterne, which has a wonderful native name impossible to write down; but, as I said before, I do not like the rather rough flavour. We had not a great variety of fruit at dessert: indeed, Sydney oranges constituted its main feature, as it is too late for winter fruits, and too early for summer ones: but we were not inclined to be over-fastidious, and thought everything delicious.

Letter III: On to New Zealand.

Christchurch, Canterbury, N. Z. October 14th, 1865. As you so particularly desired me when we parted to tell you *everything*, I must resume my story where in my last letter I left it off. If I remember rightly, I ended with an attempt at describing our great feast. We embarked the next day, and as soon as we were out of the bay the little *Albion* plunged into heavy seas. The motion was much worse in her than on board the large vessel we had been so glad to leave, and all my previous sufferings seemed insignificant compared with what I endured in my small and wretchedly hard berth. I have a dim recollection of F— — helping me to dress, wrapping me up in various shawls, and half carrying me up the companion ladder; I crawled into a sunny corner among the boxes of oranges with which the deck was crowded, and there I lay helpless and utterly miserable. One well-meaning and good-natured fellow-passenger asked F— — if I was fond of birds, and on his saying "Yes," went off for a large wicker cage of hideous "laughing Jackasses," which he was taking as a great treasure to Canterbury. Why they should be called "Jackasses" I never could discover; but the creatures certainly do utter by fits and starts a sound which may fairly be described as laughter. These paroxysms arise from no cause that one can perceive; one bird begins, and all the others join in, and a more doleful and depressing chorus I never heard: early in the morning seemed the favourite time for this discordant mirth. Their owner also possessed a cockatoo with a great musical reputation, but I never heard it get beyond the first bar of "Come into the garden, Maud." Ill as I was, I remember being roused to something like a flicker of animation when I was shown an exceedingly seedy and shabby-looking blackbird with a broken leg in splints, which its master (the same bird-fancying gentleman) assured me he had bought in Melbourne as a great bargain for only 2 pounds 10 shillings!

After five days' steaming we arrived in the open roadstead of Hokitika, on the west coast of the middle island of New Zealand, and five minutes after the anchor was down a little tug came alongside to take away our steerage passengers—three hundred diggers. The gold-fields on this coast were only discovered eight months ago, and already several canvas towns have sprung up; there are thirty thousand diggers at work, and every vessel brings a fresh cargo of stalwart, sun-burnt men. It was rather late, and getting dark, but still I could distinctly see the picturesque tents in the deep mountain gorge, their white shapes dotted here and there as far back from the shore as my sight could follow, and the wreaths of smoke curling up in all directions from the evening fires: it is still bitterly cold at night, being very early spring. The river Hokitika washes down with every fresh such quantities of sand, that a bar is continually forming in this roadstead, and though only vessels of the least possible draught are engaged in the coasting-trade, still wrecks are of frequent occurrence. We ought to have landed our thousands of oranges here, but this work was necessarily deferred till the morning, for it was as much as they could do to get all the diggers and their belongings safely ashore before dark; in the middle of the night one of the sudden and furious gales common to these seas sprang up, and would soon have driven us on the rocks if we had not got our steam up quickly and struggled out to sea, oranges and all, and away to Nelson, on the north coast of the same island. Here we landed the seventh day after leaving Melbourne, and spent a few hours wandering about on shore. It is a lovely little town, as I saw it that spring morning, with hills running down almost to the water's edge, and small wooden houses with gables and verandahs, half buried in creepers, built up the sides of the steep slopes. It was a true New Zealand day, still and bright, a delicious invigorating freshness in the air, without the least chill, the sky of a more than Italian blue, the ranges of mountains in the distance covered with snow, and standing out, sharp and clear against this lovely glowing heaven. The town itself, I must say, seemed very dull and stagnant, with little sign of life or activity about it; but nothing can be prettier or more picturesque than its situation—not unlike that of a Swiss village. Our day came to an end all too soon, and we re-embarked for Wellington, the most southern town of the North Island. The seat of government is there, and it is supposed to be a very thriving place, but is not nearly so well situated as Nelson nor so attractive to strangers. We landed and walked about a good deal, and saw what little there was to see. At first I thought the shops very handsome, but I found, rather to my disgust, that generally the fine, imposing frontage was all a sham; the actual building was only a little but at the back, looking all the meaner for the contrast to the cornices and show windows in front. You cannot think how odd it was to turn a corner and see that the building was only one board in thickness, and scarcely more substantial than the scenes at a theatre. We lunched at the principal hotel, where F— — was much amused at my astonishment at colonial prices. We had two dozen very nice little oysters, and he had a glass of porter: for this modest repast we paid eleven shillings!

We slept on board, had another walk on shore after breakfast the following morning, and about twelve o'clock set off for Lyttleton, the final end of our voyaging, which we reached in about twenty hours.

The scenery is very beautiful all along the coast, but the navigation is both dangerous and difficult. It was exceedingly cold, and Lyttleton did not look very inviting; we could not get in at all near the landing-place, and had to pay 2 pounds to be rowed ashore in an open boat with our luggage. I assure you it was a very "bad quarter of an hour" we passed in that boat; getting into it was difficult enough. The spray dashed over us every minute, and by the time we landed we were quite drenched, but a good fire at the hotel and a capital lunch soon made us all right again; besides, in the delight of being actually at the end of our voyage no annoyance or discomfort was worth a moment's thought. F— — had a couple of hours' work rushing backwards and forwards to the Custom House, clearing our luggage, and arranging for some sort of conveyance to take us over the hills. The great tunnel through these "Port Hills" (which divide Lyttleton from Christchurch, the capital of Canterbury) is only half finished, but it seems wonderful that so expensive and difficult an engineering work could be undertaken by such an infant colony.

At last a sort of shabby waggonette was forthcoming, and about three o'clock we started from Lyttleton, and almost immediately began to ascend the zig-zag. It was a tremendous pull for the poor horses, who however never flinched; at the steepest pinch the gentlemen were requested to get out and walk, which they did, and at length we reached the top. It was worth all the bad road to look down on the land-locked bay, with the little patches of cultivation, a few houses nestling in pretty recesses. The town of Lyttleton seemed much more imposing and important as we rose above it: fifteen years ago a few sheds received the "Pilgrims," as the first comers are always called. I like the name; it is so pretty and suggestive. By the way, I am told that these four ships, sent out with the pilgrims by the Canterbury Association, sailed together from England, parted company almost directly, and arrived in Lyttleton (then called Port Cooper) four months afterwards, on the same day, having all experienced fine weather, but never having sighted each other once.

As soon as we reached the top of the hill the driver looked to the harness of his horses, put on a very powerful double break, and we began the descent, which, I must say, I thought we took much too quickly, especially as at every turn of the road some little anecdote was forthcoming of an upset or accident; however, I would not show the least alarm, and we were soon rattling along the Sumner Road, by the sea-shore, passing every now and then under tremendous overhanging crags. In half an hour we reached Sumner itself, where we stopped for a few moments to change horses. There is an inn and a village here, where people from Christchurch come in the warm weather for sea-air and bathing. It began to rain hard, and the rest of the journey, some seven or eight miles, was disagreeable enough; but it was the *end*, and that one thought was sufficient to keep us radiantly good-humoured, in spite of all little trials. When we reached Christchurch, we drove at once to a sort of boarding-house where we had engaged apartments, and thought of nothing but supper and bed.

The next day people began calling, and certainly I cannot complain of any coldness or want of welcome to my new home. I like what I have seen of my future acquaintances very much. Of course there is a very practical style and tone over everything, though outwardly the place is as civilized as if it were a hundred years old; well-paved streets, gas lamps, and even drinking fountains and pillar post-offices! I often find myself wondering whether the ladies here are at all like what our great grandmothers were. I suspect they are, for they appear to possess an amount of useful practical knowledge which is quite astonishing, and yet know how to surround themselves, according to their means and opportunities, with the refinements and elegancies of life. I feel quite ashamed of my own utter ignorance on every subject, and am determined to set to work directly and learn: at all events I shall have plenty of instructresses. Christchurch is a very pretty little town, still primitive enough to be picturesque, and yet very thriving: capital shops, where everything may be bought; churches, public buildings, a very handsome club-house, etc. Most of the houses are of wood, but when they are burned down (which is often the case) they are now rebuilt of brick or stone, so that the new ones are nearly all of these more solid materials. I am disappointed to find that, the cathedral, of which I had heard so much, has not progressed beyond the foundations, which cost 8,000 pounds: all the works have been stopped, and certainly there is not much to show for so large a sum, but labour is very dear. Christchurch is a great deal more lively and bustling than most English country towns, and I am much struck by the healthy appearance of the people. There are no paupers to be seen; every one seems well fed and well clothed; the children are really splendid. Of course, as might be expected, there is a great deal of independence in bearing and manner, especially among the servants, and I hear astounding stories concerning them on all sides. My next letter will be from the country, as we have accepted an invitation to pay a visit of six weeks or so to a station in the north of the province.

Letter IV: First introduction to "Station life."

Heathstock, Canterbury, November 13th, 1865. I have just had the happiness of receiving my first budget of English letters; and no one can imagine how a satisfactory home letter satisfies the hunger of the heart after its loved and left ones. Your letter was particularly pleasant, because I could perceive, as I held the paper in my hands, that you were writing as you really felt, and that you were indeed happy. May you long continue so, dearest.

F— — says that this beautiful place will give me a very erroneous impression of station life, and that I shall probably expect to find its comforts and luxuries the rule, whereas they are the exception; in the mean time, however, I am enjoying them thoroughly. The house is only sixty-five miles from Christchurch, nearly due north (which you must not forget answers to your south in point of warmth). Our kind friends and hosts, the L— —s, called for us in their comfortable and large break, with four horses. Mr. L— — drove, F— — sat on the box, and inside were the ladies, children, and a nurse. Our first stage was to Kaiapoi, a little town on the river Waimakiriri, where we had a good luncheon of whitebait, and rested and fed the horses. From the window of the hotel I saw a few groups of Maories; they looked very ugly and peaceable, with a rude sort of basket made of flax fibres, or buckets filled with whitebait, which they wanted us to buy. There are some reserved lands near Kaiapoi where they have a very thriving settlement, living in perfect peace and good-will with their white neighbours. When we set off again on our journey, we passed a little school-house for their children.

We reached Leathfield that evening, only twenty-five miles from Christchurch; found a nice inn, or accommodation-house, as roadside inns are called here; had a capital supper and comfortable beds, and were up and off again at daylight the next morning. As far as the Weka Pass, where we stopped for dinner, the roads were very good, but after that we got more among the hills and off the usual track, and there were many sharp turns and steep pinches; but Mr. L— — is an excellent whip, and took great care of us. We all got very weary towards the end of this second day's journey, and the last two hours of it were in heavy rain; it was growing very dark when we reached the gate, and heard the welcome sound of gravel under the wheels. I could just perceive that we had entered a plantation, the first trees since we left Christchurch. Nothing seems so wonderful to me as the utter treelessness of the vast Canterbury plains; occasionally you pass a few Ti-ti palms (ordinarily called cabbage-trees), or a large prickly bush which goes by the name of "wild Irishman," but for miles and miles you see nothing but flat ground or slightly undulating downs of yellow tussocks, the tall native grass. It has the colour and appearance of hay, but serves as shelter for a delicious undergrowth of short sweet herbage, upon which the sheep live, and horses also do very well on it, keeping in good working condition, quite unlike their puffy, fat state on English pasture.

We drove through the plantation and another gate, and drew up at the door of a very large, handsome, brick house, with projecting gables and a verandah. The older I grow the more convinced I am that contrast is everything in this world; and nothing I can write can give you any idea of the delightful change from the bleak country we had been slowly travelling through in pouring rain, to the warmth and brightness of this charming house. There were blazing fires ready to welcome us, and I feel sure you will sufficiently appreciate this fact when I tell you that by the time the coal reaches this, it costs nine pounds per ton. It is possible to get Australian coal at about half the price, but it is not nearly as good.

We were so tired that we were only fit for the lowest phase of human enjoyment—warmth, food, and sleep; but the next morning was bright and lovely, and I was up and out in the verandah as early as possible. I found myself saying constantly, in a sort of ecstasy, "How I wish they could see this in England!" and not only see but feel it, for the very breath one draws on such a morning is a happiness; the air is so light and yet balmy, it seems to heal the lungs as you inhale it. The verandah is covered with honeysuckles and other creepers, and the gable end of the house where the bow-window of the drawing-room projects, is one mass of yellow Banksia roses in full blossom. A stream runs through the grounds, fringed with weeping willows, which are in their greatest beauty at this time of year, with their soft, feathery foliage of the tenderest green. The flower beds are dotted about the lawn, which surrounds the house and slopes away from it, and they are brilliant patches of colour, gay with verbenas, geraniums, and petunias. Here and there clumps of tall trees rise above the shrubs, and as a background there is a thick plantation of red and blue gums, to shelter the garden from the strong N.W. winds. Then, in front, the country stretches away in undulating downs to a chain of high hills in the distance: every now and then there is a deep gap in these, through which you see magnificent snow-covered mountains.

The inside of the house is as charming as the outside, and the perfection of comfort; but I am perpetually wondering how all the furniture—especially the fragile part of it—got here. When I remember the jolts, and ruts, and roughnesses of the road, I find myself looking at the pier-glass and glass shades, picture-frames, etc., with a sort of respect, due to them for having survived so many dangers.

The first two or three days we enjoyed ourselves in a thoroughly lazy manner; the garden was a never-ending source of delight, and there were all the animals to make friends with, "mobs" of horses to look at, rabbits, poultry, and pets of all sorts. About a week after our arrival, some more gentlemen came, and then we had a series of picnics. As these are quite unlike your highly civilized entertainments which go by the same name, I must describe one to you.

The first thing after breakfast was to collect all the provisions, and pack them in a sort of washing-basket, and then we started in an American waggon drawn by a pair of stout cobs. We drove for some miles till we came to the edge of one of the high terraces common to New Zealand scenery: here we all got out; the gentlemen unharnessed and tethered the horses, so that they could feed about comfortably, and then we scrambled down the deep slope, at the bottom of which ran a wide shallow creek. It was no easy matter to get the basket down here, I assure you; we ladies were only permitted to load ourselves, one with a little kettle, and the other with a tea-pot, but this was quite enough, as crossing the creek by a series of jumps from one wet stone to another is not easy for a beginner.

Mr. L— — brought a large dog with him, a kangaroo-hound (not unlike a lurcher in appearance), to hunt the wekas. I had heard at night the peculiar cry or call of these birds, but had not seen one until to-day. "Fly" put up several, one after another, and soon ran them down. At first I thought it very cruel to destroy such a tame and apparently harmless creature, but I am assured that they are most mischievous, and that it would be useless to turn out the pheasants and partridges which Mr. L— — has brought from England, until the numbers of the wekas are considerably reduced. They are very like a hen pheasant without the long tail feathers, and until you examine them you cannot tell they have no wings, though there is a sort of small pinion among the feathers, with a claw at the end of it. They run very swiftly, availing themselves cleverly of the least bit of cover; but when you hear a short sharp cry, it is a sign that the poor weka is nearly done, and the next thing you see is Fly shaking a bundle of brown feathers vehemently. All the dogs are trained to hunt these birds, as they are a great torment, sucking eggs and killing chickens; but still I could not help feeling sorry when Fly, having disposed of the mother, returned to the flax-bush out of which he had started her, and killed several baby-wekas by successive taps of his paw.

I have wandered away from my account of the picnic in the most unjustifiable manner. The gentlemen were toiling up the hill, after we had crossed the creek, carrying the big basket by turns between them; it was really hard work, and I must tell you in confidence, that I don't believe they liked it—at least I can answer for one. I laughed at them for not enjoying their task, and assured them that I was looking forward with pleasure to washing up the plates and dishes after our luncheon; but I found that they had all been obliged, in the early days of the colony, to work at domestic drudgery in grim and grimy earnest, so it had lost the charm of novelty which it still possessed for me.

As soon as we reached a pretty sheltered spot half-way up the hill among some trees and ferns, and by the side of the creek, we unpacked the basket, and began collecting dry wood for a fire: we soon had a splendid blaze under the lee of a fine rock, and there we boiled our kettle and our potatoes. The next thing was to find a deep hole in the creek, so over-shadowed by rocks and trees that the water would be icy cold: in this we put the champagne to cool. The result of all our preparations was a capital luncheon, eaten in a most romantic spot, with a lovely view before us, and the creek just like a Scotch burn, hurrying and tumbling down the hill-side to join the broader stream in the valley. After luncheon, the gentlemen considered themselves entitled to rest, lying lazily back among the fern and smoking, whilst we ladies sat a little apart and chatted: I was busy learning to knit. Then, about five, we had the most delicious cup of tea I ever tasted, and we repacked the basket (it was very light now, I assure you), and made our way back to the top of the terrace, put the horses in again, and so home. It was a long, bright, summer holiday, and we enjoyed it thoroughly. After a voyage, such an expedition as this is full of delight; every tree and bird is a source of pleasure.

Letter V: A pastoral letter.

Heathstock, December 1st, 1865. All I can find to tell you this month is that I have seen one of the finest and best wool-sheds in the country in full work. Anything about sheep is as new to you as it is to me, so I shall begin my story at the very beginning.

I am afraid you will think us a very greedy set of people in this part of the world, for eating seems to enter so largely into my letters; but the fact is—and I may as well confess it at once—I am in a chronic state of hunger; it is the fault of the fine air and the outdoor life: and then how one sleeps at night! I don't believe you really know in England what it is to be sleepy as we feel sleepy here; and it is delightful to wake up in the morning with the sort of joyous light-heartedness which only young children have. The expedition I am going to relate may fairly be said to have begun with eating, for although we started for our twelve miles' drive over the downs immediately after an excellent and somewhat late breakfast, yet by the time we reached the Home Station we were quite ready for luncheon. All the work connected with the sheep is carried on here. The manager has a nice house; and the wool-shed, men's huts, dip, etc., are near each other. It is the busiest season of the year, and no time could be spared to prepare for us; we therefore contented ourselves with what was described to me as ordinary station fare, and I must tell you what they gave us: first, a tureen of real mutton-broth, not hot water and chopped parsley, but excel-lent thick soup, with plenty of barley and meat in it; this had much the same effect on our appetites as the famous treacle and brimstone before breakfast in "Nicholas Nickleby," so that we were only able to manage a few little sheeps' tongues, slightly pickled; and very nice *they* were; then we finished with a Devonshire junket, with clotted cream *a discretion*. Do you think we were much to be pitied?

After this repast we were obliged to rest a little before we set out for the wool-shed, which has only been lately finished, and has all the newest improvements. At first I am "free to confess" that I did not like either its sounds or sights; the other two ladies turned very pale, but I was determined to make myself bear it, and after a moment or two I found it quite possible to proceed with Mr. L— — round the "floor." There were about twenty-five shearers at work, and everything seemed to be very systematically and well arranged. Each shearer has a trap-door close to him, out of which he pushes his sheep as soon as the fleece is off, and there are little pens outside, so that the manager can notice whether the poor animal has been too much cut with the shears, or badly shorn in any other respect, and can tell exactly which shearer is to blame. Before this plan was adopted it was hopeless to try to find out who was the delinquent, for no one would acknowledge to the least snip. A good shearer can take off 120 fleeces in a day, but the average is about 80 to each man. They get one pound per hundred, and are found in everything, having as much tea and sugar, bread and mutton, as they can consume, and a cook entirely to themselves; they work at least fourteen hours out of the twenty-four, and with such a large flock as this—about 50,000—must make a good deal.

We next inspected the wool tables, to which two boys were incessantly bringing armfuls of rolled-up fleeces; these were laid on the tables before the wool-sorters, who opened them out, and pronounced in a moment to which *bin* they belonged; two or three men standing behind rolled them up again rapidly, and put them on a sort of shelf divided into compartments, which were each labelled, so that the quality and kind of wool could be told at a glance. There was a constant emptying of these bins into trucks to be carried off to the press, where we followed to see the bales packed. The fleeces are tumbled in, and a heavy screw-press forces them down till the bale—which is kept open in a large square frame—is as full as it can hold. The top of canvas is then put on, tightly sewn, four iron pins are removed and the sides of the frame fall away, disclosing a most symmetrical bale ready to be hoisted by a crane into the loft above, where it has the brand of the sheep painted on it, its weight, and to what class the wool belongs. Of course everything has to be done with great speed and system.

I was much impressed by the silence in the shed; not a sound was to be heard except the click of the shears, and the wool-sorter's decision as he flings the fleece behind him, given in one, or at most two words. I was reminded how touchingly true is that phrase, "Like as a sheep before her shearers is dumb." All the noise is *outside*; there the hubbub, and dust, and apparent confusion are great,—a constant succession of woolly sheep being brought up to fill the "skillions" (from whence the shearers take them as they want them), and the newly-shorn ones, white, clean, and bewildered-looking, being turned out after they have passed through a narrow passage, called a "race," where each sheep is branded, and has its mouth examined in order to tell its age, which is marked in a book. It was a comfort to think all their troubles were over, for a year. You can hear nothing but barking and bleating, and this goes on from early morning till dark. We peeped in at the men's huts—a long, low wooden building, with two rows of "bunks" (berths, I should call their) in one compartment, and a table with forms round it in the other, and piles of tin plates and pannikins all about. The kitchen was near, and we were just in time to see an enormous batch of bread withdrawn from a huge brick oven: the other commissariat arrangements were on the same scale. Cold tea is supplied all day long to the shearers, and they appear to consume great quantities of it.

Our last visit was to the Dip, and it was only a short one, for it seemed a cruel process; unfortunately, this fine station is in technical parlance "scabby," and although of course great precautions are taken, still some 10,000 sheep had an ominous large S on them. These poor sufferers are dragged down a plank into a great pit filled with hot water, tobacco, and sulphur, and soused over head and ears two or three times. This torture is repeated more than once.

I was very glad to get away from the Dip, and back to the manager's house, where we refreshed ourselves by a delicious cup of tea, and soon after started for a nice long drive home in the cool, clear evening air. The days are very hot, but never oppressive; and the mornings and evenings are deliciously fresh and invigorating. You can remain out late without the least danger. Malaria is unknown, and, in spite of the heavy rains, there is no such thing as damp. Our way lay through very pretty country — a series of terraces, with a range of mountains before us, with beautiful changing and softening evening tints creeping over the whole.

I am sorry to say, we leave this next week. I should like to explore a great deal more.

Letter VI: Society. — houses and servants.

Christchurch, January 1866. I am beginning to get tired of Christchurch already: but the truth is, I am not in a fair position to judge of it as a place of residence; for, living temporarily, as we do, in a sort of boarding-house, I miss the usual duties and occupations of home, and the town itself has no place of public amusement except a little theatre, to which it is much too hot to go. The last two weeks have been *the* gay ones of the whole year; the races have been going on for three days, and there have been a few balls; but as a general rule, the society may be said to be extremely stagnant. No dinner-parties are ever given — I imagine, on account of the smallness of the houses and the inefficiency of the servants; but every now and then there is an assembly ball arranged, in the same way, I believe, as at watering-places in England only, of course, on a much smaller scale. I have been at two or three of these, and noticed at each a most undue preponderance of black coats. Nearly all the ladies were married, there were very few young girls; and it would be a great improvement to the Christchurch parties if some of the pretty and partnerless groups of a London ball-room, in all their freshness of toilette, could be transferred to them. What a sensation they would make, and what terrible heart-aches among the young gentlemen would be the result of such an importation! There were the same knots of men standing together as at a London party, but I must say that, except so far as their tailor is concerned, I think we have the advantage of you, for the gentlemen lead such healthy lives that they all look more or less bronzed and stalwart—in splendid condition, not like your pale dwellers in cities; and then they come to a ball to dance, arriving early so as to secure good partners, and their great ambition appears to be to dance every dance from the first to the last. This makes it hard work for the few ladies, who are not allowed to sit down for a moment, and I have often seen a young and pretty partner obliged to divide her dances between two gentlemen.

Although it tells only against myself, I must make you laugh at an account of a snub I received at one of these balls. Early in the evening I had danced with a young gentleman whose station was a long way "up country," and who worked so hard on it that he very seldom found time for even the mild dissipations of Christchurch; he was good-looking and gentlemanly, and seemed clever and sensible, a little *brusque*, perhaps, but one soon gets used to that here. During our quadrille he confided to me that he hardly knew any ladies in the room, and that his prospects of getting any dancing were in consequence very blank. I did all I could to find partners for him, introducing him to every lady whom I knew, but it was in vain; they would have been delighted to dance with him, but their cards were filled. At the end of the evening, when I was feeling thoroughly done up, and could hardly stand up for fatigue, my poor friend came up and begged for another dance. I assured him I could scarcely stand, but when he said in a *larmoyante* voice, "I have only danced once this evening, that quadrille with you," my heart softened, and I thought I would make a great effort and try to get through one more set of Lancers; my partner seemed so grateful, that the demon of vanity, or coquetry, or whatever it is that prompts one to say absurd things induced me to fish for a compliment, and to observe, "It was not worth while taking all the trouble of riding such a distance to dance only with me, was it?" Whereupon my poor, doleful friend answered, with a deep sigh, and an accent of profound conviction, "No, indeed it was *not*!" I leave you to imagine my discomfiture; but luckily he never observed it, and I felt all the time that I richly deserved what I got, for asking such a stupid question.

The music at these balls is very bad, and though the principal room in which they are given, at the Town Hall, is large and handsome, it is poorly lighted, and the decorations are desolate in the extreme. I am afraid this is not a very inviting picture of what is almost our only opportunity of meeting together, but it is tolerably correct. Visiting appears to be the business of some people's lives, but the acquaintance does not seem to progress beyond incessant afternoon calls; we are never asked inside a house, nor, as far as I can make out, is there any private society whatever, and the public society consists, as I have said, of a ball every now and then.

My greatest interest and occupation consist in going to look at my house, which is being cut out in Christchurch, and will be drayed to our station next month, a journey of fifty miles. It is, of course, only of wood, and seems about as solid as a band-box; but I am assured by the builder that it will be a "most superior article" when it is all put together. F— — and I made the little plan of it ourselves, regulating the size of the drawing-room by the dimensions of the carpet we brought out, and I petitioned for a little bay-window, which is to be added; so on my last visit to his timber-yard, the builder said, with an air of great dignity, "Would you wish to see the *h*oriel, mum?" The doors all come ready-made from America, and most of the wood used in building is the Kauri pine from the North Island. One advantage, at all events, in having wooden houses is the extreme rapidity with which they are run up, and there are no plastered walls to need drying. For a long time we were very uncertain where, and what, we should build on our station; but only six weeks after we made up our minds, a house is almost ready for us. The boards are sawn into the requisite lengths by machinery; and all the carpentering done down here; the frame will only require to be fitted together when it reaches its destination, and it is a very good time of year for building, as the wool drays are all going back empty, and we can get them to take the loads at reduced prices; but even with this help, it is enormously expensive to move a small house fifty miles, the last fifteen over bad roads; it is collar-work for the poor horses all the way, Christchurch being only nine feet above the sea-level, while our future home in the Malvern Hills is twelve hundred.

You know we brought all our furniture out with us, and even papers for the rooms, just because we happened to have everything; but I should not recommend any one to do so, for the expense of carriage, though moderate enough by sea (in a wool ship), is enormous as soon as it reaches Lyttleton, and goods have to be dragged up country by horses or bullocks. There are very good shops where you can buy everything, and besides these there are constant sales by auction where, I am told, furniture fetches a price sometimes under its English value. House rent about Christchurch is very high. We looked at some small houses in and about the suburbs of the town, when we were undecided about our plans, and were offered the most inconvenient little dwellings, with rooms which were scarcely bigger than cupboards, for 200 pounds a year; we saw nothing at a lower price than this, and any house of a better class, standing in a nicely arranged shrubbery, is at least 300 pounds per annum. Cab-hire is another thing which seems to me disproportionately dear, as horses are very cheap; there are no small fares, half-a-crown being the lowest "legal tender" to a cabman; and I soon gave up returning visits when I found that to make a call in a Hansom three or four miles out of the little town cost one pound or one pound ten shillings, even remaining only a few minutes at the house.

All food (except mutton) appears to be as nearly as possible at London prices; but yet every one looks perfectly well-fed, and actual want is unknown. Wages of all sorts are high, and employment, a certainty. The look and bearing of the immigrants appear to alter soon after they reach the colony. Some people object to the independence of their manner, but I do not; on the contrary, I like to see the upright gait, the well-fed, healthy look, the decent clothes (even if no one touches his hat to you), instead of the half-starved, depressed appearance, and too often cringing servility of the mass of our English population. Scotchmen do particularly well out here; frugal and thrifty, hard-working and sober, it is easy to predict the future of a man of this type in a new country. Naturally, the whole tone of thought and feeling is almost exclusively practical; even in a morning visit there is no small-talk. I find no difficulty in obtaining the useful information upon domestic subjects which I so much need; for it is sad to discover, after all my house-keeping experience, that I am still perfectly ignorant. Here it is necessary to know *how* everything should be done; it is not sufficient to give an order, you must also be in a position to explain how it is to be carried out I felt quite guilty when I saw the picture in *Punch* the other day, of a young and inexperienced matron requesting her cook "not to put any lumps into the melted butter," and reflected that I did not know how lumps should be kept out; so, as I am fortunate enough to number among my new friends a lady who is as clever in these culinary details as she is bright and charming in society, I immediately went to her for a lesson in the art of making melted butter without putting lumps into it.

The great complaint, the never-ending subject of comparison and lamentation among ladies, is the utter ignorance and inefficiency of their female servants. As soon as a ship comes in it is besieged with people who want servants, but it is very rare to get one who knows how to do anything as it ought to be done. Their lack of all knowledge of the commonest domestic duties is most surprising, and makes one wonder who in England did the necessary things of daily cottage life for them, for they appear to have done nothing for themselves hitherto. As for a woman knowing how to cook, that seems the very last accomplishment they acquire; a girl will come to you as a housemaid at 25 pounds per annum, and you will find that she literally does not know how to hold her broom, and has never handled a duster. When you ask a nurse her qualifications for the care of perhaps two or three young children, you may find, on close cross-examination, that she can recollect having once or twice "held mother's baby," and that she is very firm in her determination that "you'll keep baby yourself o' nights; mem!" A perfectly inexperienced girl of this sort will ask, and get, 30 pounds or 35 pounds per annum, a cook from 35 pounds to 40 pounds; and when they go "up country," they hint plainly they shall not stay long with you, and ask higher wages, stipulating with great exactness how they are to be conveyed free of all expense to and from their place.

Then, on the other hand, I must say they work desperately hard, and very cheerfully: I am amazed how few servants are kept even in the large and better class of houses. As a general rule, they, appear willing enough to learn, and I hear no complaints of dishonesty or immorality, though many moans are made of the rapidity with which a nice tidy young woman is snapped up as a wife; but that is a complaint no one can sympathise with. On most stations a married couple is kept; the man either to act as shepherd, or to work in the garden and look after the cows, and the woman is supposed to attend to the indoor comforts of the wretched bachelor-master: but she generally requires to be taught how to bake a loaf of bread, and boil a potato, as well as how to cook mutton in the simplest form. In her own cottage at home, who did all these things for her? These incapables are generally perfectly helpless and awkward at the wash-tub; no one seems to expect servants to know their business, and it is very fortunate if they show any capability of learning.

I must end my long letter by telling you a little story of my own personal experience in the odd ways of these girls. The housemaid at the boarding-house where we have stayed since we left Heathstock is a fat, sonsy, good-natured girl, perfectly ignorant and stupid, but she has not been long in the colony, and seems willing to learn. She came to me the other day, and, without the least circumlocution or hesitation, asked me if I would lend her my riding-habit as a pattern to give the tailor; adding that she wanted my best and newest. As soon as I could speak for amazement, I naturally asked why; she said she had been given a riding-horse, that she had loaned a saddle, and bought a hat, so now she had nothing on her mind except the habit; and further added, that she intended to leave her situation the day before the races, and that it was "her fixed intent" to appear on horseback each day, and all day long, at these said races. I inquired if she knew how to ride? No; she had never mounted any animal in her life. I suggested that she had better take some lessons before her appearance in public; but she said her mistress did not like to spare her to "practise," and she stuck steadily to her point of wanting my habit as a pattern. I could not lend it to her, fortunately, for it had been sent up to the station with my saddle, etc.; so had she been killed, as I thought not at all unlikely, at least my conscience would not have reproached me for aiding and abetting her equestrian freak. I inquired from every one who went to the races if they saw or heard of any accident to a woman on horseback, and I most anxiously watched the newspapers to see if they contained any notice of the sort, but as there has been no mention of any catastrophe, I suppose she has escaped safely. Her horse must have been quieter and better broken than they generally are. F— — says that probably it was a very old "station screw." I trust so, for her sake!

Letter VII: A young colonist.—the town and its neighbourhood.

Christchurch, March 1866. I must begin my letter this mail with a piece of domestic news, and tell you of the appearance of your small nephew, now three weeks old. The youth seems inclined to adapt himself to circumstances, and to be as sturdy and independent as colonial children generally are. All my new friends and neighbours proved most kind and friendly, and were full of good offices. Once I happened to say that I did not like the food as it was cooked at the boarding-house; and the next day, and for many days after, all sorts of dainties were sent to me, prepared by hands which were as skilful on the piano, or with a pencil, as they were in handling a saucepan. New books were lent to me, and I was never allowed to be without a beautiful bouquet. One young lady used constantly to walk in to town, some two or three miles along a hot and dusty road, laden with flowers for me, just because she saw how thoroughly I enjoyed her roses and carnations. Was it not good of her?

Christchurch has relapsed into the quietude, to call it by no harsher name. The shearing is finished all over the country, and the "squatters" (as owners of sheep-stations are called) have returned to their stations to vegetate, or work, as their tastes and circumstances may dictate. Very few people live in the town except the tradespeople; the professional men prefer little villas two or three miles off. These houses stand in grounds of their own, and form a very pretty approach to Christchurch, extending a few miles on all sides: There are large trees bordering most of the streets, which give a very necessary shade in summer; they are nearly all English sorts, and have only been planted within a few years. Poplars, willows, and the blue gum grow quickest, are least affected by the high winds, and are therefore the most popular. The banks of the pretty little river Avon, upon which Christchurch is built, are thickly fringed with weeping willows, interspersed with a few other trees, and with clumps of tohi, which is exactly like the Pampas grass you know so well in English shrubberies. I don't think I have ever told you that it has been found necessary here to legislate against water-cress. It was introduced a few years since, and has spread so rapidly as to become a perfect nuisance, choking every ditch in the neighbourhood of Christchurch, blocking up mill-streams, causing meadows to be flooded, and doing all kinds of mischief.

Towards Riccarton, about four miles out of town, the Avon shows like a slender stream a few inches wide, moving sluggishly between thick beds of water-cress, which at this time of year are a mass of white blossom. It looks so perfectly solid that whenever I am at Ilam, an insane desire to step on it comes over me, much to F— —'s alarm, who says he is afraid to let me out of his sight, lest I should attempt to do so. I have only seen one native "bush" or forest yet, and that is at Riccarton. This patch of tall, gaunt pines serves as a landmark for miles. Riccarton is one of the oldest farms in the colony, and I am told it possesses a beautiful garden. I can only see the gable-end of a house peeping out from among the trees as I pass. This bush is most carefully preserved, but I believe that every high wind injures it.

Christchurch is very prettily situated; for although it stands on a perfectly flat plain, towards the sea there are the Port Hills, and the town itself is picturesque, owing to the quantities of trees and the irregular form of the wooden houses; and as a background we have the most magnificent chain of mountains—the back-bone of the island—running from north to south, the highest peaks nearly always covered with snow, even after such a hot summer as this has been. The climate is now delicious, answering in time of year to your September; but we have far more enjoyable weather than your autumns can boast of. If the atmosphere were no older than the date of the settlement of the colony, it could not feel more *youthful*, it is so light and bright, and exhilarating! The one drawback, and the only one, is the north-west wind; and the worst of it is, that it blows very often from this point. However, I am assured that I have not yet seen either a "howling nor'-wester," nor its exact antithesis, "a sutherly buster."

We have lately been deprived of the amusement of going to see our house during the process of cutting it out, as it has passed that stage, and has been packed on drays and sent to the station, with two or three men to put it up. It was preceded by two dray-loads of small rough-hewn stone piles, which are first let into the ground six or eight feet apart: the foundation joists rest on these, so as just to keep the flooring from touching the earth. I did not like this plan (which is the usual one) at all, as it seemed to me so insecure for the house to rest only on these stones. I told the builder that I feared a strong "nor'-wester" (and I hear they are particularly strong in the Malvern Hills) would blow the whole affair away. He did not scout the idea as much as I could have wished, but held out hopes to me that the roof would "kep it down." I shall never dare to trust the baby out of my sight, lest he should be blown away; and I have a plan for securing his cradle, by putting large heavy stones in it, somewhere out of his way, so that he need not be hurt by them. Some of the houses are built of "cob," especially those erected in the very early days, when sawn timber was rare and valuable: this material is simply wet clay with chopped tussocks stamped in. It makes very thick walls, and they possess the great advantage of being cool in summer and warm in winter. Whilst the house is new nothing can be nicer; but, in a few years, the hot winds dry up the clay so much, that it becomes quite pulverized; and a lady who lives in one of these houses told me, that during a high wind she had often seen the dust from the walls blowing in clouds about the rooms, despite of the canvas and paper, and with all the windows carefully closed.

Next week F— — is going up to the station, to unpack and arrange a little, and baby and I are going to be taken care of at Ilam, the most charming place I have yet seen. I am looking forward to my visit there with great pleasure.

Letter VIII: Pleasant days at Ilam.

Ilam, April 1866. We leave this to-morrow for the station in the most extraordinary conveyance you ever saw. Imagine a flat tray with two low seats in it, perched on four very high wheels, quite innocent of any step or means of clambering in and out, and drawn, tandem-fashion, by two stout mares; one of which has a little foal by her side. The advantage of this vehicle is that it is very light, and holds a good deal of luggage. We hope to accomplish the distance — fifty miles — in a day, easily.

Although this is not my first visit to Ilam, I don't think I have ever described it to you. The house is of wood, two storeys high, and came out from England! It is built on a brick foundation, which is quite unusual here. Inside, it is exactly like a most charming English house, and when I first stood in the drawing-room it was difficult to believe: that I was at the other end of the world. All the newest books, papers, and periodicals covered the tables, the newest music lay on the piano, whilst a profusion of English greenhouse flowers in Minton's loveliest vases added to the illusion. The Avon winds through the grounds, which are very pretty, and are laid out in the English fashion; but in spite of the lawn with its croquet-hoops and sticks, and the beds of flowers in all their late summer beauty, there is a certain absence of the stiffness and trimness of English pleasure-grounds, which shows that you have escaped from the region of conventionalities. There are thick clumps of plantations, which have grown luxuriantly, and look as if they had always been there. A curve of the opposite bank is a dense mass of native flax bushes, with their tall spikes of red blossom filling the air with a scent of honey, and attracting all the bees in the neighbourhood. Ti-ti palms are dotted here and there, and give a foreign and tropical appearance to the whole. There is a large kitchen garden and orchard, with none of the restrictions of high walls and locked gates which fence your English peaches and apricots.

The following is our receipt for killing time at Ilam: — After breakfast, take the last *Cornhill* or *Macmillan,* put on a shady hat, and sit or saunter by the river-side under the trees, gathering any very tempting peach or apricot or plum or pear, until luncheon; same thing until five o'clock tea; then cross the river by a rustic bridge, ascend some turf steps to a large terrace-like meadow, sheltered from the north-west winds by a thick belt of firs, blue gums, and poplars, and play croquet on turf as level as a billiard-table until dinner. At these games the cockatoo always assists, making himself very busy, waddling after his mistress all over the field, and climbing up her mallet whenever he has an opportunity. "Dr. Lindley" — so called from his taste for pulling flowers to pieces — apparently for botanical purposes — is the tamest and most affectionate of birds, and I do not believe he ever bit any one in his life; he will allow himself to be pulled about, turned upside down, scratched under his wings, all with the greatest indifference, or rather with the most positive enjoyment. One evening I could not play croquet for laughing at his antics. He took a sudden dislike to a little rough terrier, and hunted him fairly off the ground at last, chasing him all about, barking at him, and digging his beak into the poor dog's paw. But the "Doctor's" best performance is when he imitates a hawk. He reserves this fine piece of acting until his mistress is feeding her poultry; then, when all the hens and chickens, turkeys, and pigeons are in the quiet enjoyment of their breakfast or supper, the peculiar shrill cry of a hawk is heard overhead, and the Doctor is seen circling in the air, uttering a scream occasionally. The fowls never find out that it is a hoax, but run to shelter, cackling in the greatest alarm — hens clucking loudly for their chicks, turkeys crouching under the bushes, the pigeons taking refuge in their house; as soon as the ground is quite clear, Cocky changes his wild note for peals of laughter from a high tree, and finally alighting on the top of a hen-coop filled with trembling chickens, remarks in a suffocated voice, "You'll be the death of me."

I must reverse the proverb about the ridiculous and the sublime, and finish my letter by telling you of Ilam's chief outdoor charm: from all parts of the garden and grounds I can feast my eyes on the glorious chain of mountains which I have before told you of, and my bedroom window has a perfect panoramic view of them. I watch them under all their changes of tint, and find each new phase the most beautiful. In the very early morning I have often stood shivering at my window to see the noble outline gradually assuming shape, and finally standing out sharp and clear against a dazzling sky; then, as the sun rises, the softest rose-coloured and golden tints touch the highest peaks, the shadows deepening by the contrast. Before a "nor'-wester" the colours over these mountains and in the sky are quite indescribable; no one but Turner could venture upon such a mixture of pale sea-green with deep turquoise blue, purple with crimson and orange. One morning an arch-like appearance in the clouds over the furthest ranges was pointed out to me as the sure forerunner of a violent gale from the north-west, and the prognostic was fulfilled. It was formed of clouds of the deepest and richest colours; within its curve lay a bare expanse of a wonderful green tint, crossed by the snowy *silhouette* of the Southern Alps. A few hours afterwards the mountains were quite hidden by mist, and a furious gale of hot wind was shaking the house as if it must carry it off into the sky; it blew so continuously that the trees and shrubs never seemed to rise for a moment against it.

These hot winds affect infants and children a good deal, and my baby is not at all well. However, his doctor thinks the change to the station will set him all right again, so we are hurrying off much sooner than our kind friends here wish, and long before the little house in the hills can possibly be made comfortable, though F— — is working very hard to get things settled for us.

Letter IX: Death in our new home — New Zealand children.

Broomielaw, Malvern Hills, May 1866. I do not like to allow the first Panama steamer to go without a line from me: this is the only letter I shall attempt, and it will be but a short and sad one, for we are still in the first bitterness of grief for the loss of our dear little baby. After I last wrote to you he became very ill, but we hoped that his malady was only caused by the unhealthiness of Christchurch during the autumn, and that he would soon revive and get on well in this pure, beautiful mountain air. We consequently hurried here as soon as ever we could get into the house, and whilst the carpenters were still in it. Indeed, there was only one bedroom ready for us when I arrived. The poor little man rallied at first amazingly; the weather was exquisitely bright and sunny, and yet bracing. Baby was to be kept in the open air as much as possible, so F— — and I spent our days out on the downs near the house, carrying our little treasure by turns: but all our care was fruitless: he got another and more violent attack about a fortnight ago, and after a few hours of suffering he was taken to the land where pain is unknown. During the last twelve hours of his life, as I sat before the fire with him on my lap, poor F— —kneeling in a perfect agony of grief by my side, my greatest comfort was in looking at that exquisite photograph from Kehren's picture of the "Good Shepherd," which hangs over my bedroom mantelpiece, and thinking that our sweet little lamb would soon be folded in those Divine, all-embracing Arms. It is not a common picture; and the expression of the Saviour's face is most beautiful, full of such immense feminine compassion and tenderness that it makes me feel more vividly, "In all our sorrows He is afflicted." In such a grief as this I find the conviction of the reality and depth of the Divine sympathy is my only true comfort; the tenderest human love falls short of the feeling that, without any words to express our sorrow, God knows all about it; that He would not willingly afflict or grieve us, and that therefore the anguish which wrings our hearts is absolutely necessary in some mysterious way for our highest good. I fear I have often thought lightly of others' trouble in the loss of so young a child; but now I know what it is. Does it not seem strange and sad, that this little house in a distant, lonely spot, no sooner becomes a home than it is baptized, as it were, with tears? No doubt there are bright and happy days in store for us yet, but these first ones here have been sadly darkened by this shadow of death. Inanimate things have such a terrible power to wound one: though everything which would remind me of Baby has been carefully removed and hidden away by F— —'s orders, still now and then I come across some trifle belonging to him, and, as Miss Ingelow says—

"My old sorrow wakes and cries."

Our loss is one too common out here, I am told: infants born in Christchurch during the autumn very often die. Owing to the flatness of the site of the town, it is almost impossible to get a proper system of drainage; and the arrangements seem very bad, if you are to judge from the evil smells which are abroad in the evening. Children who are born on a station, or taken there as soon as possible, almost invariably thrive, but babies are very difficult to rear in the towns. If they get over the first year, they do well; and I cannot really call to mind a single sickly, or even delicate-looking child among the swarms which one sees everywhere.

I cannot say that I think colonial children prepossessing in either manners or appearance, in spite of their ruddy cheeks and sturdy limbs. Even quite little things are pert and independent, and give me the idea of being very much spoiled. When you reflect on the utter absence of any one who can really be called a nurse, this is not to be wondered at. The mothers are thoroughly domestic and devoted to their home duties, far more so than the generality of the same class at home. An English lady, with even an extremely moderate income, would look upon her colonial sister as very hard-worked indeed. The children cannot be entrusted entirely to the care of an ignorant girl, and the poor mother has them with her all day long; if she goes out to pay visits (the only recognized social duty here), she has to take the elder children with her, but this early introduction into society does not appear to polish the young visitors' manners in the least. There is not much rest at night for the mater-familias with the inevitable baby, and it is of course very difficult for her to be correcting small delinquents all day long; so they grow up with what manners nature gives them. There seems to me, however, to be a greater amount of real domestic happiness out here than at home: perhaps the want of places of public amusement may have something to do with this desirable state of affairs, but the homes seem to be thoroughly happy ones. A married man is an object of envy to his less fortunate brethren, and he appears anxious to show that he appreciates his good fortune. As for scandal, in the ordinary acceptation of the word, it is unknown; gossip there is in plenty, but it generally refers to each other's pecuniary arrangements or trifling peculiarities, and is all harmless enough. I really believe that the life most people lead here is as simple and innocent as can well be imagined. Each family is occupied in providing for its own little daily wants and cares, which supplies the mind and body with healthy and legitimate employment, and yet, as my experience tells me, they have plenty of leisure to do a kind turn for a neighbour. This is the bright side of colonial life, and there is more to be said in its praise; but the counterbalancing drawback is, that the people seem gradually to lose the sense of larger and wider interests; they have little time to keep pace with the general questions of the day, and anything like sympathy or intellectual appreciation is very rare. I meet accomplished people, but seldom well-read ones; there is also too much talk about money: "where the treasure is, there will the heart be also;" and the incessant financial discussions are wearisome, at least to me.

Letter X: Our station home.

Broomielaw, July 1866. We are now in mid-winter, and a more delicious season cannot well be imagined; the early mornings and evenings and the nights are very cold, but the hours from 10 A.M. till 5 P.M. are exquisitely bright, and quite warm. We are glad of a fire at breakfast, which is tolerably early, but we let it out and never think of relighting it until dark. Above all, it is calm: I congratulate myself daily on the stillness of the atmosphere, but F— — laughs and says, "Wait until the spring." I bask all day in the verandah, carrying my books and work there soon after breakfast; as soon as the sun goes down, however, it becomes very cold. In an English house you would hardly feel it, but with only one plank an inch thick, a lining-board and canvas and paper, between you and a hard frost, a good fire is wanted. We burn coal found twelve miles from this; it is not very good, being only what is called "lignite." I don't know if that conveys to you a distinct impression of what it really is. I should say it was a better sort of turf: it smoulders just in the same way, and if not disturbed will remain many hours alight; it requires a log of dry wood with it to make a really good blaze. Fuel is most difficult to get here, and very expensive, as we have no available "bush" on the Run; so we have first to take out a licence for cutting wood in the Government bush, then to employ men to cut it, and hire a drayman who possesses a team of bullocks and a dray of his own, to fetch it to us: he can only take two journeys a day, as he has four miles to travel each way, so that by the time the wood is stacked it costs us at least thirty shillings a cord, and then there is the labour of sawing and cutting it up. The coal costs us one pound a ton at the mouth of the pit, and the carriage exactly doubles its price; besides which it is impossible to get more, than a small quantity at a time, on account of the effect of the atmosphere on it. Exposure to the air causes it to crumble into dust, and although we keep our supply in a little shed for the purpose, it is wasted to the extent of at least a quarter of each load. We are unusually unfortunate in the matter of firing; most stations have a bush near to the homestead, or greater facilities for draying than we possess.

You tell me to describe my little house to you, so I must try to make you see it, only prefacing my attempt by warning you not to be disgusted or disappointed at any shortcomings. The house has not been built in a pretty situation, as many other things had to be considered before a picturesque site: first it was necessary to build on a flat (as the valleys here are called), not too far off the main track, on account of having to make the road to it ourselves; the next thing to be thought of was shelter from the north-west wind; then the soil must be fit for a garden, and a good creek, or brook, which would not go dry in the summer, close at hand. At present, everything out of doors is so unfinished that the place looks rather desolate, and it will be some years before our plantations can attain a respectable size, even allowing for the rapid growth in this climate. The first step is to obtain shelter from our enemy the "nor'-wester," and for this purpose we have planted quantities of broom in all directions; even the large beds for vegetables in the garden have a hedge of Cape broom on the exposed side; fortunately, the broom grows very quickly in spite of the wind, and attains to a luxuriant beauty rarely seen in England. We have put in many other trees, such as oaks, maples, etc., but not one is higher than this table, except a few poplars; the ground immediately outside the house has been dug up, and is awaiting the spring to be sown with English grass; we have no attempt at a flower-garden yet, but have devoted our energies to the vegetable one,—putting in fruit trees, preparing strawberry and asparagus beds, and other useful things. Out of doors matters would not even be as far advanced towards a garden and plantation as they are if we had commenced operations ourselves, but the ground has been worked since last year. I am glad we have chosen to build our house here instead of at the homestead two miles off; for I like to be removed from the immediate neighbourhood of all the work of the station, especially from that of the "gallows,"—a high wooden frame from which the carcases of the butchered sheep dangle; under the present arrangement the shepherd brings us over our mutton as we want it.

Inside the house everything is comfortable and pretty, and, above all things, looks thoroughly home-like. Out of the verandah you pass through a little hall hung with whips and sticks, spurs and hats, and with a bookcase full of novels at one end of it, into a dining-room, large enough for us, with more books in every available corner, the prints you know so well on the walls, and a trophy of Indian swords and hunting-spears over the fireplace: this leads into the drawing-room, a bright, cheery little room—more books and pictures, and a writing-table in the "*h*oriel." In that tall, white, classical-shaped vase of Minton's which you helped me to choose is the most beautiful bouquet, made entirely of ferns; it is a constant object for my walks up the gullies, exploring little patches of bush to search for the ferns, which grow abundantly under their shelter by the creek. I have a small but comfortable bedroom, and there is a little dressing-room for F— — and the tiniest spare room you ever saw; it really is not bigger than the cabin of a ship. I think the kitchen is the chief glory of the house, boasting a "Leamington range" a luxury quite unknown in these parts, where all the cooking is done on an American stove,—a very good thing in its way, but requiring to be constantly attended to. There is a good-sized storeroom, in which F— — has just finished putting me up some cupboards, and a servants' room. It is not a palace is it? But it is quite large enough to hold a great deal of happiness. Outside, the premises are still more diminutive; a little wash-house stands near the kitchen door, and further up the enclosure is a stable, and a small room next it for saddles, and a fowl-house and pig-stye, and a coal-shed. Now you know everything about my surroundings; but—there is always a *but* in everything—I have one great grievance, and I hope you will appreciate its magnitude.

It was impossible for F— — to come up here when the house was first commenced, and the wretch of a builder deliberately put the drawing-and dining-room fireplaces in the corner, right up against the partition wall, of course utterly destroying the comfort as well as the symmetry of the rooms. I am convinced some economy of bricks is at the bottom of this arrangement, especially as the house was built by contract; but the builder pretends to be surprised that I don't admire it, and says, "Why, it's so oncommon, mum!" I assure you, when I first saw the ridiculous appearance of the drawing-room pier-glass in the corner, I should liked to have screamed out at the builder (like the Queen in "Alice in Wonderland"), "Cut off his head!"

When we were packing up the things to come here, our friends expressed their astonishment at our taking so many of the little elegancies of life, such as drawing-room ornaments, pictures, etc. Now it is a great mistake not to bring such things, at all events a few of them, for they are not to be bought here, and they give the new home a certain likeness to the old one which is always delightful. I do not advise people to make large purchases of elegancies for a colonial life, but a few pretty little trifles will greatly improve the look of even a New Zealand up-country drawing-room.

You have asked me also about our wardrobes. Gentlemen wear just what they would on a Scotch or English farm; in summer they require perhaps a lighter hat, and long rides are always taken in boots and breeches. A lady wears exactly what would be suitable in the country in England, except that I should advise her to eschew muslin; the country outside the home paddock is too rough for thin material; she also wants thick boots if she is a good walker, and I find nails or little screws in the soles a great help for hill-walking. A hat is my only difficulty: you really want a shady hat for a protection against the sun, but there are very few days in the year on which you can ride in anything but a close, small hat, with hardly any brim at all, and even this must have capabilities of being firmly fastened on the head. My nice, wide-brimmed Leghorn hangs idly in the hall: there is hardly a morning still enough to induce me to put it on even to go and feed my chickens or potter about the garden. This being winter, I live in a short linsey dress, which is just right as to warmth, and not heavy. It is a mistake to bring too much: a year's supply will be quite enough; fresh material can easily be procured in Christchurch or any of the large towns, or sent out by friends. I find my sewing-machine the greatest possible comfort, and as time passes on and my clothes need remodelling it will be still more use ful. Hitherto I have used it chiefly for my friends' benefit; whilst I was in town I constantly had little frocks brought to me to tuck, and here I employ it in making quilted cloth hats for my gentlemen neighbours.

Letter XI: Housekeeping, and other matters.

Broomielaw, September 1866. I am writing to you at the end of a fortnight of very hard work, for I have just gone through my first experience in changing servants; those I brought up with me four months ago were nice, tidy girls and as a natural consequence of these attractive qualities they have both left me to be married. I sent them down to Christchurch in the dray, and made arrangements for two more servants to return in the same conveyance at the end of a week. In the meantime we had to do everything for ourselves, and on the whole we found this picnic life great fun. The household consists, besides F— — and me, of a cadet, as they are called—he is a clergyman's son learning sheep-farming under our auspices—and a boy who milks the cows and does odd jobs out of doors. We were all equally ignorant of practical cookery, so the chief responsibility rested on my shoulders, and cost me some very anxious moments, I assure you, for a cookery-book is after all but a broken reed to lean on in a real emergency; it starts by assuming that its unhappy student possesses a knowledge of at least the rudiments of the art, whereas it ought not to disdain to tell you whether the water in which potatoes are to be boiled should be hot or cold. I must confess that some of my earliest efforts were both curious and nasty, but E ate my numerous failures with the greatest good-humour; the only thing at which he made a wry face was some soup into which a large lump of washing-soda had mysteriously conveyed itself; and I also had to undergo a good deal of "chaff" about my first omelette, which was of the size and consistency of a roly-poly pudding. Next to these failures I think the bread was my greatest misfortune; it went wrong from the first. One night I had prepared the tin dish full of flour, made a hole in the midst of the soft white heap, and was about to pour in a cupful of yeast to be mixed with warm water (you see I know all about it in theory), when a sudden panic seized me, and I was afraid to draw the cork of the large champagne bottle full of yeast, which appeared to be very much "up." In this dilemma I went for F— —. You must know that he possesses such extraordinary and revolutionary theories on the subject of cooking, that I am obliged to banish him from the kitchen altogether, but on this occasion I thought I should be glad of his assistance. He came with the greatest alacrity; assured me he knew all about it, seized the big bottle, shook it violently, and twitched out the cork: there was a report like a pistol-shot, and all my beautiful yeast flew up to the ceiling of the kitchen, descending in a shower on my head; and F— — turned the bottle upside down over the flour, emptying the dregs of the hops and potatoes into my unfortunate bread. However, I did not despair, but mixed it up according to the directions given, and placed it on the stove; but, as it turned out, in too warm a situation, for when I went early the next morning to look at it, I found a very dry and crusty mass. Still, nothing daunted, I persevered in the attempt, added more flour and water, and finally made it up into loaves, which I deposited in the oven. That bread *never* baked! I tried it with a knife in the orthodox manner, always to find that it was raw inside. The crust gradually became several inches thick, but the inside remained damp, and turned quite black at last; I baked it until midnight, and then I gave it up and retired to bed in deep disgust. I had no more yeast and could not try again, so we lived on biscuits and potatoes till the dray returned at the end of the week, bringing, however, only one servant. Owing to some confusion in the drayman's arrangements, the cook had been left behind, and "Meary," the new arrival, professed her willingness to supply her place; but on trial being made of her abilities, she proved to be quite as inexperienced as I was; and to each dish I proposed she should attempt, the unvarying answer was, "The missis did all that where I come from." During the first few days after her arrival her chief employment was examining the various knick-knacks about the drawing-room; in her own department she was greatly taken with the little cottage mangle. She mangled her own apron about twenty times a day, and after each attempt I found her contemplating it with her head on one side, and saying to herself, "'Deed, thin, it's as smooth as smooth; how iver does it do it?" A few days later the cook arrived. She is not all I could wish, being also Irish, and having the most extraordinary notions of the use, or rather the abuse, of the various kitchen implements: for instance, she will poke the fire with the toasting fork, and disregards my gentle hints about the poker; but at all events she can both roast mutton and bake bread. "Meary" has been induced to wash her face and braid up her beautiful hair, and now shines forth as a very pretty good-humoured girl. She is as clever and quick as possible, and will in time be a capital housemaid. She has taken it into her head that she would like to be a "first-rater," as she calls it, and works desperately hard in the prosecution of her new fancy.

I have never told you of the Sunday services we established here from the first week of our arrival. There is no church nearer than those in Christchurch, nor — I may mention parenthetically — is there a doctor within the same distance. As soon as our chairs and tables were in their proper places, we invited our shepherds and those neighbours immediately around us to attend service on Sunday afternoon at three o'clock. F— — officiates as clergyman; *my* duties resemble those of a beadle, as I have to arrange the congregation in their places, see that they have Prayer-books, etc. Whenever we go out for a ride, we turn our horses' heads up some beautiful valley, or deep gorge of a river, in search of the huts of our neighbours' shepherds, that we may tell the men of these services and invite them to attend. As yet, we have met with no refusals, but it will give you an idea of the scantiness of our population when I tell you that, after all our exertions, the "outsiders" only amount to fourteen, and of these at least half are gentlemen from neighbouring stations. With this number, in addition to our own small group, we consider that we form quite a respectable gathering. The congregation all arrive on horseback, each attended by at least two big colley dogs; the horses are turned into the paddock, the saddles deposited in the back verandah, and the dogs lie quietly down by their respective masters' equipments until they are ready to start homewards. There is something very wild and touching in these Sunday services. If the weather is quite clear and warm, they are held in the verandah; but unless it is a very sunny afternoon, it is too early in the year yet for this.

The shepherds are a very fine class of men as a rule, and I find them most intelligent; they lead solitary lives, and are fond of reading; and as I am anxious to substitute a better sort of literature in their huts than the tattered yellow volumes which generally form their scanty library, I lend them books from my own small collection. But, as I foresee that this supply will soon be exhausted, we have started a Book Club, and sent to London for twenty pounds' worth of books as a first instalment. We shall get them second-hand from a large library, so I hope to receive a good boxful. The club consists of twenty-eight members now, and will probably amount to thirty-two, which is wonderful for this district. At the close of a year from the first distribution of the books they are to be divided into lots as near as possible in value to a pound each, the parcels to be numbered, and corresponding figures written on slips of paper, which are to be shaken up in a hat and drawn at random, each member claiming the parcel of which the number answers to that on his ticket. This is the fairest way I can think of for the distribution, and every one seems satisfied with the scheme. The most popular books are those of travel or adventure; unless a novel is really very good indeed, they do not care about it.

The last little item of home news with which I must close this month's budget is, that F— — has been away for a few days on a skating excursion. A rather distant neighbour of ours called on his way up to the station far back among the hills, and gave such a glowing account of the condition of the ice in that part of the country, that F— —, who is very fond of the amusement, was persuaded to accompany him. Our friend is the son of the Bishop, and owns a large station about twenty-six miles from this. At the back of his run the hills rise to a great height, and nestled among them lie a chain of lakes, after the largest of which (Lake Coleridge) Mr. H— —'s station is named. On one of the smaller lakes, called by the classical name of "Ida," the ice attains to a great thickness; for it is surrounded by such lofty hills that during the winter months the sun hardly touches it, and it is commonly reported that a heavily-laden bullock-dray could cross it in perfect safety. F— —was away nearly a week, and appears to have enjoyed himself thoroughly, though it will seem to you more of hard work than amusement; for he and Mr. H— —, and some other gentlemen who were staying there, used to mount directly after breakfast, with their skates tied to their saddle-bow, and ride twelve miles to Lake Ida, skate all through the short winter's day, lunching at the solitary hut of a gentleman-farmer close by the lake, and when it grew dusk riding home again. The gentlemen in this country are in such good training through constant exercise, that they appear able to stand any amount of fatigue without minding it.

Letter XII: My first expedition.

Broomielaw, October 1866. This ought to be early spring, but the weather is really colder and more disagreeable than any which winter brought us; and, proverbially fickle as spring sunshine and showers are in England, ours is a far more capricious and trying season. Twice during this month have I been a victim to these sudden changes of climate; on the first occasion it was most fortunate that we had reached the shelter of a friendly and hospitable roof, for it was three days before we could re-cross the mountain-pass which lay between us and home. One beautiful spring morning F— — asked me if I would like to ride across the hills, and pay my first visit to some kind and old friends of his, who were among the earliest arrivals in the province, and who have made a lovely home for themselves at the foot of a great Bush on the other side of our range. I was delighted at the idea, for I have had very little opportunity of going about since we came here, owing to the short winter days and the amount of occupation at home consequent on a new establishment.

Directly after breakfast, the horses were caught and saddled, and we started in high spirits. As we rode up the long, sunny valley stretching away for miles at the back of the house, F— — pointed out to me, with all a sheep-farmer's pride, the hundreds of pretty little curly-fleeced lambs skipping about the low hill-sides. After we passed our own boundary fence we came upon a very bad track,— this is the name by which all roads are called, and they do not deserve a better,—but it was the only path to our destination. The air was mild and balmy, and the sun shone brightly as we slowly picked our way across bogs and creeks, and up and down steep, slippery hill-sides; but just as we reached the lowest saddle of the range and prepared to descend, a cold wind met us. In an instant the sunshine was overclouded, and F— —, pointing to a grey bank of cloud moving quickly towards us, said, "There is a tremendous sou'-wester coming up; we had better push on for shelter, or you'll be drowned:" but, alas! at each step the road grew worse and worse; where it was level the ground was literally honeycombed with deep holes half full of water, and at last we came to a place where the horse had to descend a flight of stone steps, each step being extremely slippery and some way below the other; and at the bottom of this horrible staircase there was a wide jump to be taken, the spring being off the lowest step, and the jump upwards alighting on a steep bank up which the horses scrambled like cats. Getting wet through appeared to me a very minor evil compared to the dangers of such a road, but F— — urged me forward, with assurances that the horse knew the path perfectly well and could carry me at a gallop quite safely; but it was impossible to infuse sufficient courage into my drooping heart to induce me to go faster than a walk.

All this time the storm drew rapidly nearer, the wind blew in icy cold gusts, the hail came down in large stones, pelting our faces till they tingled again; it was nearly an hour before we rode up to the hospitable, ever-open porch door of Rockwood. I was immediately lifted off my saddle by kind and strong arms, and carried with frozen limbs and streaming habit into the kitchen, for I was as unfit for the drawing-room as my own water-spaniel. A blazing wood fire was hastily lighted in one of the bed-rooms, and thither the good hostess conveyed me. I emerged from that apartment the most extraordinary figure you ever saw. Imagine me arrayed in a short and very wide crinoline, over which was a bright-coloured linsey petticoat; an old pilot-coat for a jacket, huge carpet slippers on my feet, and my dripping hair hanging loose over my shoulders! I assure you, I looked like the portraits in books of travel, of the Tahitian women when they first assumed clothes; and the worst of it was, that I had to remain in this costume for three whole days. To return was impossible, the storm from the S.W. raged all that evening. When we opened our eyes next morning, snow was lying some inches deep, and still falling fast; there was no cessation for forty-eight hours, and then we had to give it time to thaw a little, so that it was Sunday morning before we started on our homeward ride. In the meantime, nothing could afford a greater contrast to the wild weather out of doors than the snug brightness within. Blazing logs of pine and black birch made every room warm and cheery; all day we chatted and amused ourselves in different ways (I learned to make a capital pudding, and acquainted myself with the mysteries of "junket"); in the evenings we had whist for an hour, and then either round games or songs. The young men of the house have nice voices and a great feeling for music, and some of the trios and glees went very well indeed. The only thing which spoilt my enjoyment was the constantly recurring remembrance of that terrible road. F— — tried to comfort me by assurances that the snow would have filled up the worst places so much that I should not see them, but, strange to say, I failed to derive any consolation from that idea; however, we accomplished the journey back safely, but with many slips and slides. As soon as we came on our own run, F— —began to look out for dead lambs, but fortunately there were not many for him to mourn over; they must have taken shelter under the low hills, to leeward of the storm.

The second ride was much longer, and if possible a more disagreeable one. It began just in the same way; we were again decoyed out by sunshine and soft air for a ride round the run, starting about half-past ten. The scenery was beautiful, and we enjoyed ourselves immensely. The track lay along our own boundary fence most of the way, and we had ridden about ten miles, when we stopped at one of our shepherds' huts, technically called an out station, and accepted his offer of luncheon. He gave us capital tea, with an egg beaten up in it as a substitute for milk, cold mutton, bread, and a cake; the reason of these unwonted luxuries was that he kept fowls, and I was very jealous at seeing two broods of chickens out, whilst mine are still in the shell. This man is quite an artist, and the walls of his but were covered with bold pen-and-ink sketches, chiefly reminiscences of the hunting-field in England, or his own adventures "getting out" wild cattle on the Black Hills in the north of the province: he leads an extremely-solitary existence, his dogs being his only companions; his duties consist in riding daily a boundary down the gorge of the river, which he has to cross and re-cross many times: and he has to supply the home station and our house with mutton, killing four or five sheep a week. He is employed out of doors all day, but has plenty of time in the evenings for reading I found him well-informed and intelligent, and he expresses himself exceedingly well. We rested here an hour, and as we went outside and prepared to mount, F— — said, "I really believe there is *another* sou'-wester coming up," and so there was: we could not go fast, for we were riding over a dry river-bed, composed entirely of loose large stones. Every few hundred yards we had to cross the river Selwyn, which was rising rapidly, as the storm had been raging in the mountains long before it reached us; on each side were high, steep hills, and in some places the river filled up the gorge entirely, and we had to ride in the water up to our saddle-girths. All this time the rain was coming down in sheets, but the wind grew colder and colder; at last the rain turned into snow, which speedily changed us and our horses into white moving figures. Eight long weary miles of this had we, only able to trot the last two, and those over very swampy ground. In your country a severe cold would probably have been the least evil of this escapade, but here no such consequence follow a good wetting; the houses are so little real protection from the weather, that you are forced to live as it were in the open air, whether you like it or not, and this hardens the constitution so much, that it is not easy to take cold from a little extra exposure. Men are apt to be careless and remain in their wet things, or stand before a fire till their clothes dry on them; and whenever I scold any one for being foolish, he always acknowledges that if he does but change when he comes into a house, he *never* catches cold from any amount of exposure to the severest weather.

Letter XIII: Bachelor hospitality.—a gale on shore.

Broomielaw, November 1866. We have lately made a much longer excursion than those I told you of last, month, and this time have been fortunate in meeting with fine weather above all, our expedition has been over perfectly level ground, and on a good "track," which has greatly increased its charms in my eyes. A fortnight ago early summer set fairly in, and some bachelor neighbours took advantage of the change to ride over to see us, and arrange a plan for the following week. It all fitted in nicely, for F— — was obliged to go to Christchurch at that time, and the first idea of the expedition originated in my saying how dull I was at the station when he was away. I can get on very well all day; with my various employments—feeding the chickens, taking the big dogs out for a walk, and so on: but after the house is quiet and silent for the night, and the servants have gone to bed, a horrible lonely eerie feeling comes over me; the solitude is so dreary, and the silence so intense, only broken occasionally by the wild, melancholy cry of the weka. However, I am very rarely tried in this way, and when I am it can't be helped, if that is any consolation.

I forget whether I told you that we left all "evening things," and other toilette necessaries which would not be wanted up country, behind us in Christchurch, so as to avoid the trouble of sending any luggage backwards or forwards. It is necessary to mention this, to account for the very light marching order in which we travelled. It was a lovely summer morning on which we left home, meaning to be away nearly a week, from Monday till Saturday. We were well mounted, and all our luggage consisted of my little travelling-bag fastened to the pommel of my saddle, containing our brushes and combs, and what is termed a "swag" in front of F— —'s saddle; that is, a long narrow bundle, in this instance enclosed in a neat waterproof case, and fastened with two straps to the "D's," which are steel loops let in in four places to all colonial saddles, for the purpose of carrying blankets, etc.; they derive their name apparently from their resemblance to the letter. In this parcel our most indispensable garments were tightly packed. We cantered gaily along on the way to Christchurch, the horses appearing to enjoy the delicious air and soft springy turf as much as we did. There was a river and half-a-dozen creeks to be crossed; but they are all quite low at this time of year. As we stood in one of them to let the horses drink and cool their legs, I saw a huge eel hidden under the shadow of a high overhanging bank, waiting till the evening to come out and feed upon the myriads of flies and little white moths that skim over the surface of the water.

It is considered a great advantage to our station that there is only the river Selwyn (of which the Maori name is the Wai-kiri-kiri) between us and town, not only for our own convenience, but because it is easy to take sheep across it, and it offers no difficulties to the wool drays. This river has a very good reputation, and is very rarely dangerous to cross; whereas the Rakaia and the Rangitata towards the south, and the Waimakiriri towards the north, of Christchurch, are most difficult, and always liable to sudden freshes. The general mode of crossing the larger rivers is by a boat, with the horse swimming behind; but accidents constantly occur from the foolhardiness of people attempting to ford them alone on horseback: they are lost in quicksands, or carried down by the current, before they can even realize that they are in danger. The common saying in New Zealand is, that people only die from drowning and drunkenness. I am afraid the former is generally the result of the latter.

From the first our road lay with our backs to the hills; but as we cantered along the plains, I was often obliged to turn round and admire their grand outlines. The highest ranges were still snow-white, and made a magnificent background against the summer sky. An easy twelve miles' ride brought us to a charming little station, called by the pretty native name of Waireka; here lived our three bachelor hosts, and a nicer or more comfortable home in a distant land could not be desired. The house has been built for some years, consequently the plantations about it and the garden have grown up well, and the willows, gum-trees, and poplars shelter it perfectly, besides giving it such a snug home look. It stands on a vast plain, without even an undulation of the ground near it; but the mountains form a grand panoramic view. There is a large wide verandah round two sides of the house, with French windows opening into it; and I could not help feeling impatient to see my own creepers in such luxuriant, beauty as these roses and honey-suckles were. It was half amusing and half pathetic to notice the preparations which had been made to receive a lady guest, and the great anxiety of my hosts to ensure my being quite as comfortable as I am at home. Much had been said beforehand about the necessity of making up my mind to rough it in bachelor quarters, so I was surprised to find all sorts of luxuries in my room, especially a dainty little toilette-table, draped with white cloths (a big wooden packing-case was its foundation). Its ornaments were all sorts of nondescript treasures, placed in boxes at the last moment of leaving the English hall or rectory by careful loving hands of mothers and sisters, and lying unused for years until now. There was a little china tray, which had been slipped into some corner by a child-sister anxious to send some possession of her "very own" out to the other end of the world; there was a vase with flowers; a parti-coloured pincushion of very gay silks, probably the parting gift of an old nurse; and a curious old-fashioned essence bottle, with eau-de-cologne; the surrounding country had been ransacked to procure a piece of scented soap. The only thing to remind me that I was not in an English cottage was the opossum rug with which the neat little bed was covered. The sitting-room looked the picture of cosy comfort, with its well-filled book-shelves, arm-chairs, sofa with another opossum rug thrown over it, and the open fireplace filled with ferns and tufts of the white feathery Tohi grass in front of the green background. We enjoyed our luncheon, or rather early dinner, immensely after our ride; and in the afternoon went out to see the nice large garden (such a contrast to our wretched little beginnings), and finally strolled on to the inevitable wool-shed, where the gentlemen had an animated "sheep talk." I rather enjoy these discussions, though they are prefaced by an apology for "talking shop;" but it amuses me, and I like to see the samples of wool, which are generally handed about in the heat of a great argument, the long white locks are so glistening, and soft, and crinkly.

My five-o'clock tea was duly remembered, and then, as there was nothing more to see out of doors within a short distance, I proposed that I should make a cake. The necessary ingredients were quickly collected. I had relays of volunteers to beat up the eggs, and though I suffered great anxiety until it was cut at supper, it turned out satisfactorily. The worst of my cookery is, that while I always follow the same directions most carefully, there is great uncertainty and variety about the result. In the evening we played round games. But we all went early to bed, as, we had to be up betimes, and in the saddle by seven o'clock, to catch the 9-30 train at Rolleston; twenty miles off. We had a beautiful, still morning for our ride, and reached the station—a shed standing out on the plain—in time to see our horses safely paddocked before the train started for Christchurch. The distance by rail was only fifteen miles, so we were not long about it; and we walked to the hotel from the railway-station in the town. A bath and breakfast were both very enjoyable, and then F— — went out to transact his business, and I employed myself in unpacking and *ironing* a ball-dress for a party, to which we were engaged that evening. There was also another ball the following night. The second was a very late one, and we had scarcely an hour's sleep before we were obliged to get up and start by the 6 A.M. train back to Rolleston, where we remounted our horses and rode to dear little Waireka in time for breakfast. By the evening I was sufficiently rested to make another cake, which also, happily, turned out well.

We intended to return home the next day (Friday), but a terrific "nor'-wester" came on in the night, and it was impossible to stir out of the house; it was the severest gale since our arrival, and it is hardly possible to give you a correct idea of the force and fury of the wind. Not a glimpse of the mountains was to be seen; a haze of dust, as thick as any fog, shut everything out. The sheep had all taken refuge under the high banks of the creeks. It is curious that sheep always feed head to wind in a nor'-west gale, whereas they will drift for miles before a sou'-wester. The trees bent almost flat before the hot breath of this hurricane, and although the house was built of cob, and its walls were very thick and solid, the creaking and swaying of the shingled roof kept me in perpetual alarm. The verandah was a great protection; and yet the small river-pebbles, of which the garden-walk was made, were dashed against the windows like hailstones by each gust. We amused ourselves indoors by the study and composition of acrostics, and so got through an imprisonment of two days, without a moment's cessation of the wind; but towards sunset on Saturday there were signs of a lull, and about midnight the gale dropped; and we heard the grateful, refreshing sound of soft and continuous rain, and when we came out to breakfast on Sunday morning everything looked revived again. It is a most fortunate meteorological fact that these very high winds are generally succeeded by heavy rain; everything is so parched and shrivelled up by them that I do not know what would become of the vegetation otherwise. We held a council, to determine what had better be done about returning home, and finally decided to risk a wet ride sooner than disappoint the little congregation; for should it prove a fine afternoon, those who lived near would certainly come; so we mounted after breakfast.

I was wrapped in one of the gentlemen's macintoshes, and found the ride far from disagreeable. As we neared our own station we began to look out for signs of disaster; and about half a mile from the house saw some of the vanes from the chimneys on the track; a little nearer home, across the path lay a large zinc chimney-pot; then another; and when we came close enough to see the house distinctly, it looked very much dwarfed without its chimneys. There had been a large pile of empty boxes at the back of the stable; these were all blown away in the gale. One huge packing-case was sailing tranquilly about on the pond, and planks and fragments of zinc were strewn over the paddock. The moment we reached the house, Mr. U— —, the gentleman-cadet of whom I have told you, came out, with a melancholy face, to tell me that a large wooden cage, full of the canaries which I had brought from England with me, had been blown out of the verandah, though it was on the most sheltered side of the house. It really seemed incredible at first, but the cage was lying in ruins in the middle of the paddock, and all my birds except one had disappeared. It happened in the middle of the night, and Mr. U— —described, very amusingly, that when he was awakened by the noise which the cage made against a wire fence (which it just "topped" in passing), he sprang out of his bed in the attic, and clambered out of the window, expecting to find the very heavy sort of staircase-ladder in its place; but it was "over the hills and far away," so he had a drop of about twelve feet to the ground, which thoroughly aroused him. He went into the verandah to see if the cage was safe, and was nearly knocked down by a big tin bath, ordinarily kept there, which was just starting across country. As soon as he missed the cage he very pluckily went after it, being able to keep sight of it by the fitful gleams of moon-light, and he was just in time to rescue the poor little surviving canary. We could not help laughing at the recital of all the mischief which had been done, but still it is very tiresome, and the garden looks, if possible, more wretched than ever. There is no shelter for it yet, and my poor green-peas are blown nearly out of the ground. It rained hard all the evening, so our congregation was confined to the home party.

Letter XIV: A Christmas picnic, and other doings.

Broomielaw, December 1866. It is too late to wish you a merry Christmas and a happy New Year in this letter. In order to allow them to reach you in time I should have sent my good wishes in October's letter; I must remember to do so next year. I am writing on the last days of the month, so I shall be able to tell you of our own Christmas doings; though, first, I must describe the festivities attending a "coming of age in the Bush," to which we were invited about the middle of this month. How strange Christmas picnics and balls will appear in your eyes, before which still dangle probably the dear old traditional holly and ivy! I am obliged to preface all my descriptions with an account of a ride, if I am to begin, according to your repeated injunctions, at the very beginning; for a ride is quite certain to be both the beginning and end of each excursion, simply because we have no other means of going about, except on our feet. The ride upon this occasion was to Rockwood, where the birthday party was to assemble, but the road had not now so many terrors for me. In consequence of the fine dry weather, most of the bad places were safer and firmer, and the numerous creeks were only shallow sparkling streamlets over which a child could jump, instead of the muddy noisy wide brooks of three months ago. The day on which we started, this time, was a great contrast to the former one. When we reached the saddle I have before told you of, instead of being met and nearly driven back by a violent "sutherly buster," we stopped before beginning the steep descent to admire the exquisite view before us.

Close on our right hand rose the Government bush out of which we get our firewood, standing grand and gloomy amid huge cliffs and crags; even the summer sunshine could not enliven it, nor the twitter and chirrup of countless birds. In front, the chain of hills we were crossing rolled down in gradually decreasing hillocks, till they merged in the vast plains before us, stretching away as far as the eye could reach towards the south, all quivering in the haze and glare of the bright sunlight. The background, extending along the horizon, was formed of lofty mountains still glistening white against the dazzling blue sky. Just at our feet the Rockwood paddocks looked like carpets of emerald velvet, spread out among the yellowish tussocks; the fences which enclose them were either golden with broom and gorse, or gay with wild roses and honeysuckle. Beyond these we saw the bright patches of flowers in the garden, and nothing could be more effective than the white gable of the house standing out against the vast black birch forest which clothed the steep hill-sides for miles — the contrast was so picturesque between the little bit of civilization and culture and the great extent of wild, savage scenery around it. After the utter treelessness of our own immediate neighbourhood, the sight of such a mass of foliage is a joy to my eyes.

The day following our arrival was *the* birthday, and we prepared to enjoy every hour of it. The party assembled was a very large one, consisting, however, chiefly of gentlemen, for the utmost exertions in the district could not produce more than five ladies altogether, and two of those had come an immense way. Directly after breakfast we all sallied forth, the ladies equipped in light cotton dresses (muslin is too thin for the bush) and little sailor hats, — we did not want shady ones, for never a gleam of sun can penetrate into a real New Zealand Bush, unless in a spot which has been very much cleared. Strong boots with nails in the soles, to help us to keep on our feet up the steep clay hill-sides, and a stout stick, completed our equipment; perhaps we were not very smart, but we looked like going at all events. I can answer for myself that I enjoyed every moment of that long Midsummer holiday most intensely, though I fear I must have wearied our dear, charming host, by my incessant questions about the names of the trees and shrubs, and of the habits and ways of the thousands of birds. It was all so new and so delightful to me, — the green gloom, the hoarse croak of the ka-ka, as it alighted almost at our feet and prepared, quite careless of our vicinity, to tear up the loose soil at the root of a tall tree, in search of grubs. It is a species of parrot, but with very dingy reddish-brown plumage, only slightly enlivened by a few, scarlet feathers in the wing. The air was gay with bright green parroquets flitting about, very mischievous they are, I am told, taking large tithe of the fruit, especially of the cherries. Every now and then we stood, by common consent, silent and almost breathless to listen to the Bell-bird, a dingy little fellow, nearly as large as a thrush with the plumage of a chaffinch, but with such a note! — how can I make you hear its wild, sweet, plaintive tone, as a little girl of the party said, "just as if it had a bell in its throat;" but indeed it would require a whole peal of silver bells to ring such an exquisite chime. Then we crept softly up to a low branch, to have a good look at the Tui, or Parson-bird, most respectable and clerical-looking in its glossy black suit, with a singularly trim and dapper air, and white wattles of very slender feathers — indeed they are as fine as hair-curled coquettishly at each side of his throat, exactly like bands. All the birds were quite tame, and, instead of avoiding us, seemed inclined to examine us minutely. Many of them have English names, which I found very tantalising, especially when, the New Zealand Robin was announced, and I could only see a fat little ball of a bird, with a yellowish-white breast. Animals there are none. No quadruped is indigenous to New Zealand, except a rat; but then, on the other hand, we are as free from snakes and all vermin as if St. Patrick himself had lived here. Our host has turned several pheasants into this forest, but they increase very slowly on account of the wekas. However, the happiness of this morning was made complete by our putting up two splendid rocketers.

We could only make our way by the paths which have been cut through the Bush; a yard off the track it is impossible to stir for the dense undergrowth. In the ravines and steep gullies formed by the creeks grow masses of ferns of all sorts, spreading like large shrubs, and contrasting by their light bright green with the black stems of the birch-trees around them. There are a few pines in this bush, but not many. I can give you no idea of the variety among the shrubs: the koromika, like an Alpine rose, a compact ball of foliage; the lance-wood, a tall, slender stem, straight as a line, with a few long leaves at the top, turned downwards like the barb of a spear, and looking exactly like a lance stuck into the ground; the varieties of matapo, a beautiful shrub, each leaf a study, with its delicate tracery of black veins on a yellow-green ground; the mappo, the gohi, and many others, any of which would be the glory of an English shrubbery: but they seem to require the deep shelter of their native Bush, for they never flourish when transplanted. I noticed the slender the large trees have of the ground, and it is not at all surprising, after such a gale as we had three weeks ago, to see many of the finest blown down in the clearings where the wind could reach them. They do not seem to have any tap-root at all, merely a very insufficient network of fibres, seldom of any size, which spreads a short way along the surface of the ground As long as a Bush is undisturbed by civilization, it appears to be impervious to wind or weather; but as soon as it is opened and cleared a little, it begins to diminish rapidly. There are traces all over the hills of vast forests having once existed; chiefly of totara, a sort of red pine, and those about us are scattered with huge logs of this valuable wood, all bearing traces of the action of fire; but shepherds, and explorers on expeditions, looking for country, have gradually consumed them for fuel, till not many pieces remain except on the highest and most inaccessible ranges.

It was a delightful, and by no means unacceptable surprise which awaited us on the other side, when, on emerging from a very thick part of the Bush, we came on a lovely spot, a true "meeting of the waters." Three broad, bright creeks came rushing and tumbling down from the densely wooded hills about to join and flow on in quite a good-sized river, amid boulders and a great deal of hurry and fuss,—a contrast to the profound quiet of our ramble hitherto, the silence of which was only broken by the twitter and whistle of the birds. Never a song can you hear, only a sweet chirrup, or two or three melodious notes. On the opposite bank of the river there was the welcome sight of several hampers more or less unpacked, and the gleam of a white tablecloth on the moss. Half-a-dozen gentlemen had formed themselves into a commissariat, and were arranging luncheon. We could see the champagne cooling in a sort of little bay, protected by a dam of big stones from being carried down the stream. It all looked very charming and inviting, but the next question was how to get across the river to these good things. Twelve or fourteen feet separated us, hungry and tired wanderers as we were, from food and rest; the only crossing-place was some miles lower down, near the house in fact; so even the most timid amongst us scouted the idea of retracing our steps. The only alternative was to make a bridge: one of the gentlemen who were with us carried an axe in case of emergency, and in a moment we heard the sharp ringing sounds foretelling the fall of a tree. In the meantime, others of the party were dragging out fallen logs — of course small and manageable ones — and laying them from one huge boulder to another, working up to their knees in water. So many of these prostrate trunks were "convenient," that a cry soon arose to the woodman to "spare the trees," for there were quite enough on the ground. However, two substantial poles had been felled, and these were laid over the deepest and most dangerous part of the current. The bridge was soon declared passable, and loud shouts from the opposite side proclaimed that luncheon was quite ready. I was called, as having a most undeserved reputation for "pluck," to make trial of the aerial-looking fabric. I did not like it at all, and entreated some one else to lead the forlorn hope; so a very quiet young lady, who really possessed more courage in her little finger than I do in my whole body, volunteered to go first. The effect from the bank was something like tight-rope dancing, and it was very difficult to keep one's balance. Miss Kate, our pioneer, walked on very steadily, amid great applause, till she reached the middle of the stream, where fortunately the water was shallow, but strewed with masses of boulders. She paused an instant on the large rock on which the ends of the saplings rested, and then started afresh for the last half of her journey. The instant she put her foot on the second part of the bridge, it gave way with a loud crash; and the poor girl, with great presence of mind, caught at the tree she, had just crossed, and so saved herself from a ducking. Of course, she had plenty of help in an instant, but the difficulty was to regain any sort of footing. She could not drop into the water, and there was apparently no way of dragging herself up again; but one of the gentlemen crept on hands and knees along the unbroken part of the bridge, and eventually helped her up the sides of the large boulder which acted as a pier, and from which the log had slipped. From the other side they now pushed across tall, slim trees, freshly cut, and the rest of the passage was safe enough. I did not like the mode of transit at all, though I got over without a slip, but it requires a steady head to cross a noisy stream on two slippery round poles — for really the trees were little thicker — laid side by side, bending with every step. It was a great comfort to me all luncheon-time to know that we were not to return by the same path through the Bush. We had a good rest after lunch: I lay back on a bed of fern, watching the numbers of little birds around us; they boldly picked up our crumbs, without a thought of possible danger. Presently I felt a tug at the shawl on which I was lying: I was too lazy and dreamy to turn my head, so the next thing was a sharp dig on my arm, which hurt me dreadfully. I looked round, and there was a weka bent on thoroughly investigating the intruder into its domain. The bird looked so cool and unconcerned, that I had not the heart to follow my first impulse and throw my stick at it; but my forbearance was presently rewarded by a stab on the ankle, which fairly made me jump up with a scream, when my persecutor glided gracefully away among the bushes, leaving me, like Lord Ullin, "lamenting."

We sauntered home slowly, gathering armfuls of, fern and a large variety of a stag's-head moss so common on the west coast of Scotland; and as soon as we had had some tea, the gentlemen went off with their towels to bathe in the creek, and the five ladies set to work at the decorations for the ball-room, weaving wreaths and arranging enormous bouquets very rapidly: we had such a wealth of flowers to work with that our task was not difficult. The most amusing part of the story is, however, that the ball took place in my bed-room! A very pompous lady of my acquaintance always prefaces the slenderest anecdote with these words, "And it happened in this wise," so I think I shall avail myself of the *tour de phrase*.

It happened in this wise, then:-a large well-proportioned room had been added to the house lately; it was intended for a drawing-room, but for some reason has only been used as a spare bed-room, but as it may possibly return to its original destination, very little bed-room furniture has been put in it, and many of its belongings are appropriate to a sitting-room. We called in the servants, the light cane bedstead was soon deposited under the shade of a tree in the garden, the washing-stand was similarly disposed of, and an hour's work with hammer and nails and a ball of string turned the room into a perfect bower of ferns and flowers: great ingenuity was displayed in the arrangement of lights, and the result was a very pretty ball-room.

We are always eating in this country, so you will not be surprised to hear that there was yet another meal to be disposed of before we separated to dress in all sorts of nooks and corners. White muslin was the universal costume, as it can be packed flat and smooth. My gown had been carried over by F— — in front of his saddle in a very small parcel: I covered it almost entirely with sprays of the light-green stag's-head, moss, and made a wreath of it also for my hair. I think that with the other ladies roses were the most popular decoration, and they looked very fresh and nice. I was the universal *coiffeuse*, and I dressed all the girls' heads with flowers, as I was supposed to be best up in the latest fashions. In the meantime, the piano had been moved to the bay-window of the ball-room, and at ten o'clock dancing commenced, and may be truly said to have been kept up with great spirit until four o'clock: it only ceased then on account of the state of exhaustion of the unfortunate five ladies, who had been nearly killed with incessant dancing. I threw a shawl over my head, and sauntered alone up one of the many paths close to the house which led into the Bush. Tired as I was, I shall never forget the beauty and romance of that hour,—the delicious crisp *new* feeling of the morning air; the very roses, growing like a red fringe on the skirts of the great Bush, seemed awaking to fresh life and perfume; the numbers of gay lizards and flies coming out for their morning meal, and, above all, the first awakening of the myriads of Bush-birds; every conceivable twitter and chatter and chirrup; the last cry of a very pretty little owl, called, from its distinctly uttered words, the "More-pork," as it flitted away before the dawn to the highest trees: all made up a jubilant uproar compared to which one of the Crystal Palace choruses is silence. I sat down on a fallen tree, and listened and waited: every moment added to the lovely dawn around me, and I enjoyed to the full the fragrant smells and joyous sounds of another day in this fresh young land.

All too soon came a loud "coo-ee" from the house, which I allowed them to repeat before I answered; this was to tell me that the ball-room was deserted, and had been again turned into a bedroom. When I opened my eyes later, after a six hours' nap, the room looked like a fairy bower, the flowers still unfaded. We had another picnic the next day up the gorge of a river, amid very wild and beautiful scenery; but everything had been arranged so as to make the expedition an easy one, out of consideration to the weary five. The day after this we rode home again, and I had to set to work directly to prepare for my own Christmas party to the shepherds and shearers,—for we have just commenced to muster the sheep, and the shearing will be in full force by Christmas Day. One great object I have in view in giving this party is to prevent the shearers from going over to the nearest accommodation-house and getting tipsy, as they otherwise would; so I have taken care to issue my invitations early. I found great difficulty in persuading some of the men to accept, as they had not brought any tidy clothes with them; and as the others would be decently, indeed well dressed, they did not like putting in a shabby appearance. This difficulty was obviated by F— — hunting up some of the things he had worn on the voyage, and rigging-out the invited guests. For two days before the great day I had been working hard, studying recipes for pies and puddings, and scouring the country in search of delicacies. Every lady was most kind, knowing that our poor, exposed garden was backward; I had sacks of green peas, bushels of young potatoes, and baskets of strawberries and cherries sent to me from all round the country; I made poor F— — ride twenty miles to get me a sirloin of beef, and, to my great joy, two beautiful young geese arrived as a present only the day before. It is a point of honour to have as little mutton as possible on these occasions, as the great treat is the complete change of fare. I only ventured to introduce it very much disguised as curry, or in pies. We were all up at daylight on Christmas morning, and off to the nearest little copse in one of the gullies, where a few shrubs and small trees and ferns grow, to gather boughs for the decoration of the washhouse. Marvels were done in the carpentering line to arrange tables around its walls. The copper, which at first presented such an obstacle to the symmetry of the adornments, became their chief glory; it was boarded over, its sides completely hidden by flags and ferns, and the dessert placed on it peeped out from a bower of greenery. I don't know how we got our own breakfast; from eleven o'clock there was the constant announcement "A horseman coming up the flat;" and by twelve, when I as beadle announced that all was ready, a large congregation of thirty-six came trooping into my little drawing-room. As soon as it was filled the others clustered round the door; but all could hear, I think. F— — began the service; and as the notes of the Christmas Anthem swelled up, I found the tears trembling in my eyes. My overwhelming thought was that it actually was the very first time those words had ever been sung or said in that valley—you in England can hardly realize the immensity of such a thought—"the first time since the world was made." I think the next sensation was one of extreme happiness; it seemed such a privilege to be allowed to hold the initial Christmas service. I had to grasp this idea very tight to keep down the terrible home-sickness which I felt all day for almost the first time. There are moments when no advantages or privileges can repress what Aytoun calls "the deep, unutterable woe which none save exiles feel."

The service only lasted half an hour, beginning and ending with a hymn; there were three women present besides me—my two servants, and the nice young wife of a neighbouring shepherd. It was a sultry day, not a breath of air; but still it is never oppressive at this elevation. We wound up a big musical-box, set it going in the banqueting-hall (late washhouse), and marshalled the guests in they were extremely shy as a rule, and so we soon went away and left them to themselves. They ate incessantly for two hours—and I hope they enjoyed themselves; then the men lounged about the stables and smoked, and the three women cleared away a little. F— — and our gentlemen guests got up athletic sports in the shade which seemed very popular, though it appeared a great deal of trouble to take on such a hot day. As the sun sank below the hills it grew much cooler, and my two maids came with a shamefaced request to be allowed to dance in the kitchen. I inquired about the music?— that was provided for by a fiddle and some pipes; so I consented, but I found they wanted me to start them. I selected as my partner a very decent young farmer who lives near, but has left his farm and is at work branding our sheep all shearing-time. The pride and delight of his mate was much greater than my partner's; he stood near his friend, prompting him through the mazes of the most extraordinary quadrille you ever saw, with two extra figures. Then there was an endless polka, in which everybody danced, like Queen Elizabeth, "high and disposedly;" but the ball ended at nine o'clock, and we were given some cold dinner, for which we were all very ready. The next morning saw the remains of the festivity cleared away, and every one hard at work again; for this is our very busiest season. The work of the station, however, is carried on at the homestead two miles off. F— — is there all day long, but I see nothing of it. While the shearers' hearts were tender, I asked them to come over to church on Sunday, and they have promised to do so: I lend them quantities of books and papers also, so as to keep them amused and away from the accommodation-house.

Letter XV: Everyday station life.

Broomielaw, January 1867. You tell me to describe our daily home-life and domestic surroundings. I dare say it: will appear to be a monotonous and insignificant existence enough when put on paper, but it suits me exactly; and, for the first time in my life, I have enough to do, and also the satisfaction of feeling that I am of some little use to my fellow-creatures. A lady's influence out here appears to be very great, and capable of indefinite expansion. She represents refinement and culture (in Mr. Arnold's sense of the words), and her footsteps on a new soil such as this should be marked by a trail of light. Of course every improvement must be the work of time, but I find my neighbours very willing to help me in my attempts.

A few lines will be sufficient to sketch a day's routine. The first of my duties is one I especially delight in. I am out very early with a large tin dish of scraps mixed with a few handfuls of wheat, and my appearance is the signal for a great commotion among all my fowls and ducks and pigeons. Such waddling and flying and running with outstretched wings to me: in fact, I receive a morning greeting from all the live-stock about the place. I am nearly knocked down by the big sheep-dogs; the calves come rushing with awkward gambols towards me for a bit of the fowls' bread, whilst the dogs look out for a bone; but, in the midst of the confusion, the poultry hold their own; indeed, an anxious hen eager to secure a breakfast for her chicks will fly at a big dog, and beat him away from a savoury morsel. I think I ought not to omit mentioning the devotion of a small pig; it is an exact illustration of the French proverb which speaks of the inequality of love, for I am quite passive and do not respond in the least to the little beastie's affection, which is the most absurd thing you ever saw, especially as it proceeds from so unromantic an animal. Late in the spring (that is to say, about November last) we were all returning from a great pig-hunting expedition, when I saw one of the party coming down a steep hill near the house with a small and glossy-black wild pig under each arm; he was very proud of his captives, placed them in a box with some straw, and fed them like babies out of a bottle. We laughed at him very much; but when he went away he begged so earnestly that the pigs should be reared that we promised to keep them. In a few days they became perfectly tame, and were very handsome little creatures; and one of them attached itself to me, following me all about, even into the house (but *that* I really could not stand), accompanying me in all my walks, and, as far as it could, in my rides. Many a time have I seen poor little piggy carried down a creek by the current, squealing piteously, but it was evidently a case of "many waters cannot quench love," for a little further on piggy would appear, very much baked, but holding out gallantly, till sheer exhaustion compelled him to give in, when he would lie down under a tussock, apparently dying; but, as we were coming home in the dusk, Helen, my pretty bay mare, has given many a shy at piggy starting up from his shelter with gambols and squeals of joy.

It is always a great temptation to loiter about in the lovely fresh morning air, but I have to be dressed in time for prayers and breakfast at nine; directly after breakfast I go into the kitchen; sometimes, it is only necessary to give orders or instructions, but generally I find that practice is much better than precept, and I see to the soup myself, and make the pudding—the joint can take care of itself.

You have often asked me what we have to eat, so this will be a good opportunity of introducing our daily bill of fare, prefacing it with my recorded opinion that here is no place in the world where you can live so cheaply and so well as on a New Zealand sheep station, when once you get a start. Of course, it is expensive at first, setting everything going, but that would be the case in any country. I will begin at the very beginning:—Porridge for breakfast, with new milk and cream *a discretion*; to follow—mutton chops, mutton ham, or mutton curry, or broiled mutton and mushrooms, not shabby little fragments of meat broiled, but beautiful tender steaks off a leg; tea or coffee, and bread and butter, with as many new-laid eggs as we choose to consume. Then, for dinner, at half-past one, we have soup, a joint, vegetables, and a pudding; in summer, we have fresh fruit stewed, instead of a pudding, with whipped cream. I was a proud and happy woman the first day my cream remained cream, and did not turn into butter; for generally my zeal outran my discretion, and I did not know when to leave off whipping. We have supper about seven; but this is a moveable feast, consisting of tea again, mutton cooked in some form of entree, eggs, bread and butter, and a cake of my manufacture. I must, however, acknowledge, that at almost every other station you would get more dainties, such as jam and preserves of all sorts, than we can boast of yet; for, as Littimer says to David Copperfield, "We are very young, exceedingly young, sir," our fruit-trees, have not come into full bearing, and our other resources are still quite undeveloped.

However, I have wandered away terribly from my first intention of telling you of the daily occupations to a description of our daily food. After I have finished all my little fussings about the house, I join F— — who has probably been for some time quietly settled down at his writing-table, and we work together at books and writing till dinner; after that meal, F— — like Mr. Tootes, "resumes his studies," but I go and feed my fowls again, and if I am very idly disposed I sit on a hencoop in the shade and watch the various tempers of my chickens and ducklings. A little later F— — and I go out for some hours: if it is not too hot, he takes his rifle and we go over the hills pig-stalking, but this is really only suitable exercise for a fine winter's day; at this time of year we either go for a walk or a ride, generally the latter—not a little shabby canter, but a long stretching gallop for miles and miles; perhaps stopping to have a cup of tea with a neighbour twelve or fifteen miles off, and then coming slowly home in the delicious gloaming, with the peculiar fresh crisp feeling which the atmosphere always has here the moment the sun sets, no matter how hot the day has been. I can hardly hope to make you understand how enjoyable our twilight hours are, with no fear of damp or malaria to spoil them; every turn of the track as we slowly wind up the valley showing us some beautiful glimpse of distant mountain peaks, and, above all, such sunset splendours, gradually fading away into the deep, pure beauty of a summer night.

In one of our rides the other day, after crossing a low range of hills, we suddenly dropped down on what would be called in England a hamlet, but here it is designated by the extraordinary name of a "nest of cockatoos." This expression puzzled me so much when I first heard it, that I must give you as minute an explanation as I myself found necessary to the comprehension of the subject.

When a shepherd has saved a hundred pounds, or the better class of immigrant arrives with a little capital, the favourite investment is in freehold land, which they can purchase, in sections of twenty acres and upwards, at 2 pounds the acre. The next step is to build a sod but with two rooms on their property, thatching it with Tohi, or swamp grass; a door and a couple of window-frames all ready glazed are brought from Christchurch in the dray with the family and the household goods. After this rough and ready shelter is provided, the father and sons begin fencing their land and gradually it all assumes a cultivated appearance. Pig-sties and fowl-houses are added; a little garden, gay with common English flowers, is made in front of the house, whose ugly walls are gradually hidden by creepers, and the homestead looks both picturesque and prosperous. These small farmers are called Cockatoos in Australia by the squatters or sheep-farmers, who dislike them for buying up the best bits of land on their runs; and say that, like a cockatoo, the small freeholder alights on good ground, extracts all he can from it, and then flies away to "fresh fields and pastures new." But the real fact is, that the poor farmer perhaps finds his section is too far from a market, so he is forced to abandon it and move nearer a town, where the best and most productive land has been bought up already; and he has to begin again at a disadvantage. However, whether the name is just or not, it is a recognized one here; and I have heard a man say in answer to a question about his usual occupation, "I'm a Cockatoo."

This particular "nest" appeared to me very well off, comparatively speaking; for though the men complained sadly of the low price of their wheat and oats, still there was nothing like poverty to be seen. Ready money was doubtless scarce, and an extensive system of barter appeared to prevail; but still they all looked well fed and well clothed; sickness was unknown among them, and it did one's heart good to see the children—such sturdy limbs, bright fearless eyes, and glowing faces. They have abundance of excellent food. Each cottager has one or two cows, and the little ones take these out to pasture on the hills, so they are in the open air nearly all day: but their ignorance is appalling! Many of them had never even been christened; there was no school or church within thirty miles or more, and although the parents seemed all tidy, decent people, and deplored the state of things, they were powerless to help it. The father and elder sons work hard all day; the mother has to do everything, even to making the candles, for the family; there is no time or possibility of teaching the children. The neighbouring squatters do not like to encourage settlers to buy up their land, therefore they carefully avoid making things pleasant for a new "nest," and the Cockatoos are "nobody's business;" so, as far as educational advantages go, they are perfectly destitute.

When I mentioned my discovery of this hamlet, and my dismay at the state of neglect in which so many fine intelligent-looking children were growing up, every one warned me not to interfere, assuring me the Cockatoo was a very independent bird, that he considered he had left all the Ladies Bountiful and blanket and coal charities behind him in the old country; that, in short, as it is generally put, "Jack is as good as his master" out here, and any attempt at patronage would be deeply resented. But I determined to try the effect of a little visiting among the cottages, and was most agreeably surprised at the kind and cordial welcome I received. The women liked to have some one to chat to about their domestic affairs, and were most hospitable in offers of tea, etc., and everywhere invitations to "come again" were given; so the next week I ventured to invite the men over to our Sunday services. Those who were fond of reading eagerly accepted the offer to join the book-club, and at last we started the educational subject. Many plans were discussed, and finally we arranged for one woman, who had received an excellent education and was quite fitted for the post, to commence a day-school; but this entailed so much loss of her valuable time that the terms she is obliged to ask seem disproportionately high to the people's means. She wants 2 shillings and 6 pence a week with each child, and this is terrible heavy on the head of a family who is anxious and willing to give them some "schooling." However, the plan is to be tried, and I have promised to start them with books, slates, copybooks, etc. It was quite touching to hear their earnest entreaties that F— — would come over on Sunday sometimes and hold a service there, but I tried to show them this could not be managed. The tears actually came into their eyes when I talked of the happiness it would be to see a little church and school in their midst; and the almost invariable remark was, "Ah, but it'll be a far day first." And so I fear it will—a very far day; but I have often heard it said, that if you propose one definite object to yourself as the serious purpose of your life, you will accomplish it some day. Well, the purpose of my life henceforward is to raise money somehow or somewhere to build a little wooden school-room (licensed for service, to be held whenever a missionary clergyman comes by), and to pay the salary of a schoolmaster and mistress, so that the poor Cockatoo need not be charged more than threepence a week for each child. The Board of Education will give a third of the sum required, when two-thirds have been already raised; but it is difficult to collect subscriptions, or indeed to induce the squatters to listen to any plan for improving the condition of the small farmers, and every year which slips away and leaves these swarms of children in ignorance adds to the difficulty of training them. [Note: Since this was written, a school-house, also used as a church, has been built in this district by private subscription and Government aid. A clergyman, who lives some twenty-five miles away, rides over and holds service once a month.]

Letter XVI: A sailing excursion on Lake Coleridge.

Lake Coleridge, February 1867. A violent storm of wind and rain from the south-west keeps us all indoors to-day, and gives me time to write my letter for the Panama mail, which will be made up to-morrow. The post-office is ten miles off, and rejoices in the appropriate name of "Wind-whistle;" it stands at the mouth of a deep mountain gorge, and there never was such a temple of the winds.

This bad weather comes after a long spell of lovely bright summer days, and is very welcome to fill up the failing creeks in the lower ranges of hills. I must tell you how much we have been enjoying our visit here. F— — knows this part of the country well, but it is quite new to me, and a great contrast to the other scenery I have described to you We had long talked of paying Mr. C. H— — a visit at his bachelor cottage on his station far back among the high ranges of hills, but no time was fixed, so I was rather taken by surprise when last week he drove up to Broomielaw in a light American waggon with a pair of stout horses, and announced that he had come to take us to his place next day. There was no reason against this plan, and we agreed at once; the next morning saw us on the road, after an early breakfast. We had to drive about thirty-five miles round, whereas it would have been only twenty miles riding across the hills; but our kind host thought that it would be much more comfortable for me to be able to take a carpet-bag in the carriage instead of the usual system of saddle-bags one is obliged to adopt travelling on horseback. We made our first stage at the ever-hospitable station of the C— —'s, on the Horarata, but we could not remain to luncheon, as they wished, having to push on further; and, as it turned out, it was most fortunate we took advantage of the first part of the day to get over the ground between us and our destination, for the gentle breeze which had been blowing since we started gradually freshened into a tremendous "nor'-wester," right in our teeth all the rest of our way. The poor horses bent their heads as low as possible and pulled bravely at their collars, up hill the whole time. Among the mountains the wind rushed with redoubled fury down the narrow gorges, and became icily cold as we neared the snowy ranges. It was impossible to see the hills for the thick mist, though I knew we must have a magnificent view before us. We took refuge for an hour just to rest the horses, at Windwhistle, and I certainly expected the house to come down whilst we were there. I can hardly tell you anything of the rest of the drive, for I was really frightened at my first experience of a "howling nor'-wester" out of doors, and Mr. H— — made me sit down at the bottom of the carriage and heaped over me all the cloaks and shawls we had brought. It was delightful to find ourselves under shelter at last in a pretty bright snug room, with lots of books and arm-chairs, and a blazing fire; *this*, you must remember, in midsummer.

The next morning was perfectly calm, and the lake as serene as if no storm had been dashing its water in huge breakers against the beach only a few hours before. The view from the sitting-room was lovely: just beneath the window there was a little lawn, as green as possible from the spray with which the lake had washed it yesterday; beyond this a low hedge, an open meadow, a fringe of white pebbly beach, and then a wide expanse of water within one little wooded island, and shut in gradually from our view by spurs of hills running down to the shore, sometimes in bold steep cliffs, and again in gentle declivities, with little strips of bush or scrub growing in the steep gullies between them. The lake extends some way beyond where we lose sight of it, being twelve miles long and four miles broad. A few yards from the beach it is over six hundred feet deep. Nothing but a painting could give you any idea of the blue of sky and water that morning; the violent wind of yesterday seemed to have blown every cloud below the horizon, for I could not see the least white film anywhere. Behind the lower hills which surround the lake rises a splendid snowy range; altogether, you cannot imagine a more enchanting prospect than the one I stood and looked at; it made me think of Miss Procter's lines —

> "My eyes grow dim, As still I gaze and gaze Upon that mountain pass, That leads — or so it seems — To some far happy land Known in a world of dreams."

All this time, whilst I was looking out of the window in most unusual idleness, Mr. H — — and F — — were making constant journeys between the boat-house and the store-room, and at last I was entreated to go and put on my hat. While doing this I heard cupboards being opened, and a great bustle; so when I reached the shore I was not so much surprised as they expected, to see in the pretty little sailing-boat (which was moored to a primitive sort of jetty made out of a broken old punt) the materials for at least two substantial meals, in case of being kept out by a sudden head-wind. I was especially glad to notice a little kettle among the *impedimenta*, and there were cloaks and wraps of all kinds to provide against the worst. Four gentlemen and I made up the crew and passengers, and a very merry set we were, behaving extremely like children out for a holiday. The wind was a trifle light for sailing, so the gentlemen pulled, but very lazily and not at all in good "form," as the object of each oarsman seemed to be to do as little work as possible. However, we got on somehow, a light puff helping us now and then, but our progress was hardly perceptible. I had been for a long time gazing down into the clear blue depth of water, every now and then seeing a flash of the white sand shining at the bottom, when I was half startled by our host standing suddenly up in the bow of the boat; and then I found that we were a couple of miles away from our starting-point, and that we had turned a corner formed by a steep spur, and were running right into what appeared a grove of rata-trees growing at the water's edge. The rata only grows in the hills and near water; it is a species of broad-leaf myrtle, with a flower exactly like a myrtle in character, but of a brilliant deep scarlet colour, and twice as large.

When the bowsprit touched the rata-branches, which drooped like a curtain into the water, Mr. H— — made a signal to lower the mast, and parting the thick, blossom-covered foliage before us, with both hands, the way the boat had on her sent us gently through the screen of scarlet flowers and glossy green leaves into such a lovely fairy cove! Before us was a little white beach of fine sparkling sand, against which the water broke in tiny wavelets, and all around a perfect bower of every variety of fern and moss, kept green by streams no thicker than a silver thread trickling down here and there with a subdued tinkling sound. We all sat quite silent, the boat kept back just inside the entrance by the steersman holding on to a branch. It was a sudden contrast from the sparkling sunshine and brightness outside, all life and colour and warmth, to the tender, green, profound shade and quiet in this "Mossy Hum," as the people about here call it. Do not fancy anything damp or chilly. No; it was like a natural temple—perfect repose and refreshment to the eyes dazzled with the brilliant outside colouring. Centuries ago there must have been a great landslip here, for the side of the mountain is quite hollowed out, and Nature has gradually covered the ugly brown rent with the thickest tapestry of her most delicate handiwork. I noticed two varieties of the maiden-hair, its slender black stem making the most exquisite tracery among the vivid greens. There was no tint of colour except green when once we passed the red-fringed curtain of rata-branches, only the white and shining fairy beach and the gleaming threads of water. As we sat there, perfectly still, and entranced, a sort of delicious mesmeric feeling stole over me; I thought of the lotus-eater's chant, "There is no joy but calm," with, for, the first time in my life, a dim perception of what they meant, perhaps; but it was over all too quickly: prosaic words of direction to back water called us from shade to light, and in a moment more we were in front of the rata-trees, admiring their splendid colouring, and our little boat was dancing away over the bright waves, with her white wings set and her bows pointed towards the little toy island in the middle of the lake; it was no question now of rowing, a nice fresh breeze from the south (the *cold* point here) sent us swiftly and steadily through the water. What a morning it was! The air was positively intoxicating, making you feel that the mere fact of being a living creature with lungs to inhale such an atmosphere was a great boon. We have a good deal of disagreeable weather, and a small proportion of bad weather, but in no other part of the world, I believe, does Nature so thoroughly understand how to make a fine day as in New Zealand.

A little after mid-day we ran our boat to the lee of the island, and whilst she was steadied by the same primitive method of holding on to branches of manuka and other scrub, I scrambled out and up a little cliff, where a goat could hardly have found footing, till I reached a spot big enough to stand on, from whence I anxiously watched the disembarkation of some of the provisions, and of the gridiron and kettle. In a few moments we were all safely ashore, and busy collecting dry fern and brushwood for a fire; it was rather a trial of patience to wait till the great blaze had subsided before we attempted to cook our chops, which were all neatly prepared ready for us. Some large potatoes were put to bake in the ashes; the tin plates were warmed (it is a great art not to overheat them when you have to keep them on your lap whilst you eat your chop). We were all so terribly hungry that we were obliged to have a course of bread and cheese and sardines *first*; it was really quite impossible to wait patiently for the chops. The officiating cook scolded us well for our Vandalism, and the next moment we detected him in the act of devouring a half-raw potato. The fragments of our meal must have been a great boon to the colony of wekas who inhabit the island, for as they increase and multiply prodigiously their provisions must often fall short in so small a space. No one can imagine how these birds originally came here, for the island is at least two miles from the nearest point of land; they can neither swim nor fly; and as every man's hand is against them, no one would have thought it worth while to bring them over: but here they are, in spite of all the apparent impossibilities attending their arrival, more tame and impudent than ever. It was dangerous to leave your bread unwatched for an instant, and indeed I saw one gliding off with an empty sardine tin in its beak; I wonder how it liked oil and little scales. They considered a cork a great prize, and carried several off triumphantly.

After luncheon there was the usual interval of rest, and pipes on the part of the gentlemen. I explored a little, but there is nothing very pretty or abundant in the way of wild flowers in the parts of New Zealand which I have seen. White violets and a ground clematis are the only ones I have come across in any quantity. The manuka, a sort of scrub, has a pretty blossom like a diminutive Michaelmas daisy, white petals and a brown centre, with a very aromatic odour; and this little flower is succeeded by a berry with the same strong smell and taste of spice. The shepherds sometimes make an infusion of these when they are very hard-up for tea; but it must be like drinking a decoction of cloves.

About three o'clock we re-embarked, and sailed a little higher up the lake beyond the point where we lose sight of it from Mr. H— —'s house, every moment opening out fresh and more beautiful glimpses. Quite the opposite end of the shore is fringed with a thick deep forest, and another station has been built there, at which, I am told, the scenery is still more magnificent. At first I was inclined to wonder where the sheep live amid all this picturesque but mountainous country: however, I find that between and among these hills stretch immense valleys (or "flats," as they are called here), which are warm and sheltered in winter, and afford plenty of food for them; then, in summer, they go up to the mountains: but it is very difficult to "muster" these ranges. I am almost ashamed to confess to another meal before we returned home, but there was a lovely tempting spot in a little harbour, and so we landed and boiled some water and had a capital cup of tea. You require to be out as we were from morning till night in such an air as this to know what it is to feel either hungry or sleepy in perfection! The next day we made a similar excursion, exploring the opposite shore of the lake; but, before we started, our host distrusted the appearance of certain clouds, and sent round horses to meet us at the point where we were going to lunch; and it was just as well he did so, for a stiff breeze sprang up from the south-west, which would have kept us out all night. So we mounted the horses instead of re-embarking, having first secured the boat, and cantered home. We passed several smaller lakes; there is a perfect chain of them among these hills, and I was much amused at the names bestowed on them, according to the tastes or caprice of the station-owners whose runs happen to include them: for instance, two are called respectively "Geraldine" and "Ida," whilst three, which lie close together, rejoice in the somewhat extraordinary names of "the World," "the Flesh," and "the Devil."

Letter XVII: My first and last experience of "camping out."

Broomielaw, April 1867. I have nothing to tell you this mail, except of a rather ridiculous expedition which we made last week, and which involved our spending the whole night on the top of the highest hill on our run. You will probably wonder what put such an idea into our heads, so I must preface my account by a little explanation. Whenever I meet any people who came here in the very early days of the colony — only sixteen years ago, after all! — I delight in persuading them to tell me about their adventures and hardships during those primitive times, and these narratives have the greatest fascination for me, as they always end happily. No one ever seems to have died of his miseries, or even to have suffered seriously in any way from them, so I find the greatest delight in listening to the stories of the Pilgrims. I envy them dreadfully for having gone through so much with such spirit and cheerfulness, and ever since I came here I have regretted that the rapid advance of civilization in New Zealand precludes the possibility of being really uncomfortable; this makes me feel like an impostor, for I am convinced that my English friends think of me with the deepest pity, as of one cut off from the refinements and comforts of life, whereas I really am surrounded by every necessary, and many of its luxuries, and there is no reason but that of expense why one should not have all of these.

One class of narratives is peculiarly attractive to me. I like to hear of benighted or belated travellers when they have had to "camp out," as it is technically called; and have lived in constant hope of meeting with an adventure which would give me a similar experience. But I am gradually becoming convinced that this is almost impossible by fair means, so I have been trying for some time past to excite in the breasts of our home party and of our nearest neighbours an ardent desire to see the sun rise from the top of "Flagpole," a hill 3,000 feet above the level of the sea, and only a: couple of miles from the house. As soon as they were sufficiently enthusiastic on the subject, I broached my favourite project of our all going up there over-night, and camping out on the highest peak. Strange to say, the plan did not meet with any opposition, even from F— —, who has had to camp out many a winter's night, and with whom, therefore, the novelty may be said to have worn off. Two gentlemen of the proposed party were "new chums" like myself, and were strongly in favour of a little roughing; new-chums always are, I observe. F— — hesitated a little about giving a final consent on the score of its being rather too late in the year, and talked of a postponement till next summer, but we would not listen to such an idea; so he ended by entering so heartily into it, that when at last the happy day and hour came, an untoward shower had not the least effect in discouraging him.

There was a great bustle about the little homestead on that eventful Tuesday afternoon. Two very steady old horses were saddled, one for me and the other for one of the "new chums," who was not supposed to be in good form for a long walk, owing to a weak knee. Everything which we thought we could possibly want was heaped on and around us after we had mounted; the rest of the gentlemen, four in number, walked, and we reached the first stage of our expedition in about an hour. Here we dismounted, as the horses could go no further in safety. The first thing done was to see to their comfort and security; the saddles were carefully deposited under a large flax-bush in case of rain, and the long tether ropes were arranged so as to ensure plenty of good feed and water for both horses, without the possibility of the ropes becoming entangled in each other or in anything else. Then came a time of great excitement and laughing and talking, for all the "swags" had to be packed and apportioned for the very long and steep ascent before us.

And now I must tell you exactly what we took up. A pair of large double blankets to make the tent of,—that was one swag, and a very unwieldy one it was, strapped knapsack fashion, with straps of flax-leaves, on the back, and the bearer's coat and waistcoat fastened on the top of the whole. The next load consisted of one small single blanket for my sole use, inside of which was packed a cold leg of lamb. I carried the luncheon basket, also strapped on my shoulders, filled with two large bottles of cream, some tea and sugar, and, I think, teaspoons. It looked a very insignificant load by the side of the others, but I assure you I found it frightfully heavy long before I had gone half-way up the hill. The rest distributed among them a couple of large heavy axes, a small coil of rope, some bread, a cake, tin plates and pannikins, knives and forks, and a fine pigeon-pie. Concerning this pie there were two abominable propositions; one was to leave it behind, and the other was to eat it then and there: both of these suggestions were, however, indignantly rejected. I must not forget to say we included in the commissariat department two bottles of whisky, and a tiny bottle of essence of lemon, for the manufacture of toddy. We never see a real lemon, except two or three times a year when a ship arrives from the Fiji islands, and then they are sixpence or a shilling apiece. All these things were divided into two large heavy "swags," and to poor F— — was assigned the heaviest and most difficult load of all—the water. He must have suffered great anxiety all the way, for if any accident had happened to his load, he would have had to go back again to refill his big kettle; this he carried in his hand, whilst a large tin vessel with a screw lid over its mouth was strapped on his back also full of water, but he was particularly charged not to let a drop escape from the spout of the kettle; and I may mention here, that though he took a long time about it, for he could not go as straight up the hill as we did, he reached the top with the kettle full to the brim—the other vessel was of course quite safe. All these packings and repackings, and the comfortable adjustment of the "swags," occupied a long time, so it was past five when we began our climb, and half-past six when we reached the top of the hill, and getting so rapidly dark that we had to hurry our preparations for the night, though we were all so breathless that a "spell" (do you know that means *rest*?) would have been most acceptable. The ascent was very steep, and there were no sheep-tracks to guide us; our way lay through thick high flax-bushes, and we never could have got on without their help. I started with a stick, but soon threw it aside and pulled myself up by the flax, hand over hand. Of course I had to stop every now and then to rest, and once I chose the same flax-bush where three young wild pigs had retired for the night, having first made themselves the most beautiful bed of tussock grass bitten into short lengths; the tussocks are very much scattered here, so it must have been an afternoon's work for them; but the shepherds say these wild pigs make themselves a fresh bed every night.

The first thing to be done was to pitch the tent on the little flat at the very top of the hill: it was a very primitive affair; two of the thinnest and longest pieces of totara, with which Flagpole is strewed, we used for poles, fastening another piece lengthwise to these upright sticks as a roof-tree: this frame was then covered with the large double blanket, whose ends were kept down on the ground by a row of the heaviest stones to be found. The rope we had brought up served to tie the poles together at the top, and to fasten the blanket on them; but as soon as the tent had reached this stage, it was discovered that the wind blew through it from end to end, and that it afforded very little protection. We also found it much colder at the top of this hill than in our valley; so under these circumstances it became necessary to appropriate my solitary blanket to block up one end of the tent and make it more comfortable for the whole party. It was very little shelter before this was done. The next step was to collect wood for a fire, which was not difficult, for at some distant time the whole of the hill must have been covered by a forest of totara trees; it has apparently been destroyed by fire, for the huge trunks and branches which still strew the steep sides are charred and half burnt. It is a beautiful wood, with a strong aromatic odour, and blazed and crackled splendidly in the clear, cool evening air, as we piled up a huge bonfire, and put the kettle on to boil. It was quite dusk by this time, so the gentlemen worked hard at collecting a great supply of wood, as the night promised to be a very cold one, whilst I remained to watch the kettle, full of that precious liquid poor F— — had carried up with such care, and to prevent the wekas from carrying off our supper, which I had arranged just inside the tent. In this latter task I was nobly assisted by my little black terrier Dick, of whose sad fate I must tell you later.

By eight o'clock a noble pile of firewood had been collected, and we were very tired and hungry; so we all crept inside the tent, which did not afford very spacious accommodation, and began our supper. At this point of the entertainment everybody voted it a great success; although the wind was slowly rising and blowing from a cold point, and our blanket-tent did not afford the perfect warmth and shelter we had fondly credited it with. The gentlemen began to button up their coats. I had only a light serge jacket on, so I coaxed Dick to sit at my back and keep it warm; for, whilst our faces were roasted by the huge beacon-fire, there was a keen and icy draught behind us. The hot tea was a great comfort, and we enjoyed it thoroughly, and after it was over the gentlemen lit their pipes, and I told them a story: presently we had glees, but by ten o'clock there was no concealing the fact that we were all very sleepy indeed; however, we still loudly declared that camping out was the most delightful experiment. F—— and another gentleman (that kind and most good-natured Mr. U——, who lives with us) went outside the tent, armed with knives, and cut all the tussocks they could feel in the darkness, to make me a bed after the fashion of the pigs; they brought in several armfuls, and the warmest corner in the tent was heaped with them; I had my luncheon-basket for a pillow, and announced that I had turned in and was very comfortable, and that camping out was charming; the gentlemen were still cheery, though sleepy; and the last thing I remember was seeing preparations being made for what a Frenchman of my acquaintance always will call a "grogs." When I awoke, I thought I must have slept several hours. Though the fire was blazing grandly, the cold was intense: I was so stiff I could hardly move; all my limbs ached dreadfully, and my sensations altogether were new and very disagreeable. I sat up with great difficulty and many groans, and looked round: two figures were coiled up, like huge dogs, near me; two more, moody and sulky, were smoking by the fire; with their knees drawn up to their noses and their hands in their pockets, collars well up round their throats—statues of cold and disgust. To my inquiries about the hour, the answer, given in tones of the deepest despondency, was

"Only eleven o'clock, and the sun doesn't rise till six, and its going to be the coldest night we've had this year." The speaker added, "If it wasn't so dark that we'd break our necks on the way, we might go home."

Here was a pretty end to our amusement. I slowly let myself down again, and tried to go to sleep, but that relief was at an end for the night; the ground seemed to grow harder every moment, or, at all events, I ached more, and the wind certainly blew higher and keener. Dick proved himself a most selfish doggie; he would creep round to leeward of *me*, whilst I wanted him to let me get leeward of him, but he would not consent to this arrangement. Whenever I heard a deeper moan or sigh than usual, I whispered an inquiry as to the hour, but the usual reply, in the most cynical voice, was, "Oh, you need not whisper, nobody is asleep." I heard one plaintive murmur "Think of all our warm beds, and of our coming up here from choice." I must say I felt dreadfully ashamed of myself for my plan; it was impossible to express my contrition and remorse, for, always excepting Mr. U— —, they were all too cross to be spoken to. It certainly was a weary, long night. About one o'clock I pretended to want some hot tea, and the preparation for that got through half an hour, and it warmed us a little; but everybody still was deeply dejected, not to say morose. After an interval of only two hours more of thorough and intense wretchedness we had a "grogs," but there was no attempt at conviviality—subdued savageness was the prevailing state of mind. I tried to infuse a little hope into the party, by suggestions of a speedy termination to our misery, but my own private opinion was that we should all be laid up for weeks to come with illness. I allotted to myself in this imaginary distribution of ills a severe rheumatic fever; oh! how I ached, and I felt as if I never could be warm again. The fire was no use; except to afford occupation in putting on wood; it roasted a little bit of you at a time, and that bit suffered doubly from the cold when it was obliged to take its share of exposure to the wind. I cannot say whether the proverb is true of other nights, but this particular night, certainly, was both darkest and coldest just before dawn.

At last, to our deep joy, and after many false alarms, we really all agreed that there was a faint streak of grey in the east. My first impulse was to set off home, and I believe I tried to get up expressing some such intention, but F— — recalled me to myself by saying, in great surprise, "Are you not going to stop and see the sun rise?" I had quite forgotten that this was the avowed object of the expedition, but I was far too stiff to walk a yard, so I was obliged to wait to see what effect the sunrise would have on my frozen limbs, for I could not think of any higher motive. Presently some one called out "There's the sea," and so it was, as distinct as though it were not fifty miles off; none of us had seen it since we landed; to all of us it is associated with the idea of going home some day: whilst we were feasting our eyes on it a golden line seemed drawn on its horizon; it spread and spread, and as all the water became flooded with a light and glory which hardly seemed to belong to this world, the blessed sun came up to restore us all to life and warmth again. In a moment, in less than a moment, all our little privations and sufferings vanished as if they had never existed, or existed only to be laughed at. Who could think of their "Ego" in such a glorious presence, and with such a panorama before them? I did not know which side to turn to first. Behind me rose a giant forest in the far hills to the west—a deep shadow for miles, till the dark outline of the pines stood out against the dazzling snow of the mountains behind it; here the sky was still sheltering the flying night, and the white outlines looked ghostly against the dull neutral tints, though every peak was sharply and clearly defined; then I turned round to see before me such a glow of light and beauty! For an immense distance I could see the vast Canterbury plains; to the left the Waimakiriri river, flowing in many streams, "like a tangled bunch of silver ribbons" (as Mr. Butler calls it in his charming book on New Zealand), down to the sea; beyond its banks the sun shone on the windows of the houses at Oxford, thirty miles off as the crow would fly, and threw its dense bush into strong relief against the yellow plains. The Port Hills took the most lovely lights and shadows as we gazed on them; beyond them lay the hills of Akaroa, beautiful beyond the power of words to describe. Christchurch looked quite a large place from the great extent of ground it appeared to cover. We looked onto the south: there was a slight haze over the great Ellesmere Lake, the water of which is quite fresh, though only separated from the sea by a slight bar of sand; the high banks of the Rakaia made a deep dark line extending right back into the mountains, and beyond it we could see the Rangitata faintly gleaming in the distance; between us and the coast were green patches and tiny homesteads, but still few and far between; close under our feet, and looking like a thread beneath the shadow of the mountain, ran the Selwyn in a narrow gorge, and on its bank stood the shepherd's hut that I have told you once afforded us such a good luncheon; it looked a mere toy, as if it came out of a child's box of playthings, and yet so snug for all its lonely position. On the other hand lay our own little home, with the faint wreath of smoke stealing up through the calm air (for the wind had dropped at sunrise). Here and there we saw strings of sheep going down from their high camping-grounds to feed on the sunny slopes and in the warm valleys. Every moment added to our delight and enjoyment; but unfortunately it was a sort of happiness which one can neither speak of at the time, nor write about afterwards: silence is its most expressive language. Whilst I was drinking in all the glory and beauty before me, some of the others had been busy striking the tent, repacking the loads, very much lighter without the provisions; and we had one more excellent cup of tea before abandoning the encampment to the wekas, who must have breakfasted splendidly that morning. Our last act was to collect all the stones we could move into a huge cairn, which was built round a tall pole of totara; on the summit of this we tied securely, with flax, the largest and strongest pocket-handkerchief, and then, after one look round to the west — now as glowing and bright as the radiant east — we set off homewards about seven o'clock; but it was long before we reached the place where we left the horses, for the gentlemen began rolling huge rocks down the sides of the hills and watching them crashing and thundering into the valleys, sometimes striking another rock and then bounding high into the air. They were all as eager and excited as schoolboys, and I could not go on and leave them, lest I should get below them and be crushed under a small stone of twenty tons or so. I was therefore forced to keep well *above* them all the time. At last we reached the spur where the horses were tethered, re-saddled and loaded them, and arrived quite safely at home, just in time for baths and breakfast. I was amused to see that no one seemed to remember or allude to the miseries and aches of that long cold night; all were full of professions of enjoyment. But I noticed that the day was unusually quiet; the gentlemen preferred a bask in the verandah to any other amusement, and I have reason to believe they indulged in a good many naps.

Letter XVIII: A journey "down south."

Waimate, May 1867. In one of my early letters from Heathstock I told you that the Hurunui, which is the boundary of that run, marks the extreme north of the Province of Canterbury; and now I am writing to you from the extreme south. I hope you do not forget to reverse in your own mind the ordinary ideas of heat and cold, as connected with those points of the compass. The distance from our house to this is about 160 miles, and we actually took two days and a half to get here! — besides, into these miles was compressed the fatigue of a dozen English railway journeys of the same length. But, I suppose, as usual, you will not be satisfied unless I begin at the very beginning. The first difficulty was to reach the point where we were to join the coach on the Great South Road. It was less than thirty miles, so we could easily have ridden the distance; but the difficulty was to get our clothes all that way. They could not be carried on horseback, and just then the station-dray was particularly employed; besides which it would have taken three days to come and go, — rather a useless expenditure of the man's time, as well as of the horses' legs, where only two little portmanteaus were concerned. Fortunately for us, however, this is a country where each man is ready and willing to help his neighbour, without any inquiry as to who he is; so the moment our dilemma was known various plans were suggested for our assistance, of which this was the one selected: —

On a certain bright but cold Wednesday afternoon, F— — and I and our modest luggage started in a neighbour's "trap" for the station I have already mentioned on the Horarata, where Mr. C. H— — and I stopped on our way to Lake Coleridge. It is on the plains at the foot of a low range of downs, and about twelve miles from us. You cannot imagine a more charming little cottage *ornee* than the house is, capable of holding, apparently, an indefinite number of people, and with owners whose hospitality always prompts them to try its capabilities to the utmost. A creek runs near the house, and on its banks, sloping to the sun, lies a lovely garden, as trim as any English parterre, and a mass of fruit and flowers. Nothing can be more picturesque than the mixture of both. For instance, on the wall of the house is a peach-tree laden every autumn with rosy, velvet-cheeked fruit; and jasmine and passion-flowers growing luxuriantly near it. Inside all is bright neatness and such a welcome! As for our supper, on this particular day it comprised every dainty you can imagine, and made me think of my housekeeping with shame and confusion of face. We had a very merry evening, with round games; but there was a strong prejudice in favour of going to bed early, as we all had to be up by three o'clock: and so we were, to find a delicious breakfast prepared for us, which our kind hostess was quite disappointed to see we could not eat much of. Coffee and toast was all I could manage at that hour. We started in the dark, and the first thing we had to cross was a dry river-bed, in which one of the horses lay deliberately down, and refused to move. This eccentricity delayed us very much; but we got him into a better frame of mind, and accomplished our early drive of sixteen miles in safety, reaching the accommodation-house, or inn, where the coach from Christchurch to Timaru changes horses for its first stage, by six o'clock. There we had a good breakfast, and were in great "form" by the time the coach was ready to start. These conveyances have a worldwide celebrity as "Cobb's coaches," both in America and Australia, where they are invariably the pioneers of all wheeled vehicles, being better adapted to travel on a bad road, or no road at all, than any other four-wheeled "trap." They are both strong and light, with leathern springs and a powerful break; but I cannot conscientiously say they are at all handsome carriages; indeed I think them extremely ugly and not very comfortable except on the box-seat next the driver. Fortunately, this is made to hold three, so F— — and I scrambled up, and off we started with four good strong horses, bearing less harness about them than any quadrupeds I ever saw; a small collar, slender traces, and very thin reins comprised all their accoutrements. The first half of the journey was slow, but there was no jolting. The road was level, though it had not been made at all, only the tussocks removed from it; but it was naturally good—a great exception to New Zealand roads. The driver was a steady, respectable man, very intelligent; and when F— —could make him talk of his experiences in Australia in the early coaching days, I was much interested.

We crossed the Rakaia and the Rangitata in ferry-boats, and stopped on the banks of the Ashburton, to dine about one o'clock, having changed horses twice since we started from "Gigg's," as our place of junction was elegantly called. Here all my troubles began. When we came out of the little inn, much comforted and refreshed by a good dinner, I found to my regret that we were to change drivers as well as horses, and that a very popular and well known individual was to be the new coachman. As our former driver very politely assisted me to clamber up on the box-seat, he recommended F— — to sit on the outside part of the seat, and to put me next the driver, "where," he added, "the lady won't be so likely to tumble out." As I had shown no disposition to fall off the coach hitherto, I was much astonished by this precaution, but said nothing. So he was emboldened to whisper, after looking round furtively, "And you jest take and don't be afraid, marm; *he* handles the ribbings jest as well when he's had a drop too much as when he's sober, which ain't often, however." This last caution alarmed me extremely. The horses were not yet put in, nor the driver put *up*, so I begged F— — to get down and see if I could not go inside. But, after a hasty survey, he, said it was quite impossible: men smoking, children crying, and, in addition, a policeman with a lunatic in his charge, made the inside worse than the outside, especially in point of atmosphere; so he repeated the substance of our ex-driver's farewell speech; and when I saw our new charioteer emerge at last from the bar, looking only very jovial and tolerably steady as to gait, I thought perhaps my panic was premature. But, oh, what a time I had of it for nine hours afterwards! The moment the grooms let go the horses' heads he stood up on his seat, shook the reins, flourished his long whip, and with one wild yell from him we dashed down a steep cutting into the Ashburton. The water flew in spray far over our heads, and the plunge wetted me as effectually as if I had fallen into the river. I expected the front part of the coach to part from the back, on account of the enormous strain caused by dragging it over the boulders. We lurched like a boat in a heavy sea; the "insides" screamed; "Jim" (that was the driver's name) swore and yelled; the horses reared and plunged. All this time I was holding on like grim death to a light iron railing above my head, and one glance to my left showed me F— — thrown off the very small portion of cushion which fell to his share, and clinging desperately to a rude sort of lamp-frame. I speculated for an instant whether this would break; and, if so, what would become of him. But it took all my ideas to keep myself from being jerked off among the horses' heels. We dashed through the river; Jim gathered up the reins, and with a different set of oaths swore he would punish the horses for jibbing in the water. And he did punish them; he put the break hard down for some way, flogged them with all his strength, dancing about the coach-box and yelling like a madman. Every now and then, in the course of his bounds from place to place, he would come plump down on my lap; but I was too much frightened to remonstrate; indeed, we were going at such a pace against the wind, I had very little breath to spare.

We got over the first stage of twenty miles at this rate very quickly, as you may imagine; but, unfortunately, there was an accommodation-house close to the stables, and Jim had a good deal more refreshment. Strange to say, this did not make him any wilder in manner—that he could not be; but after we started again he became extremely friendly with me, addressing me invariably as "my dear," and offering to "treat me" at every inn from that to Timaru. I declined, as briefly as I could, whereupon he became extremely angry, at my doubting his pecuniary resources apparently, for, holding the reins carelessly with one hand, though we were still tearing recklessly along, he searched his pockets with the other hand, and produced from them a quantity of greasy, dirty one-pound notes, all of which he laid on my lap, saying, "There, and there, and there, if you think I'm a beggar!" I fully expected them to blow away, for I could not spare a hand to hold them; but I watched my opportunity when he was punishing the unfortunate fresh team, and pounced on them, thrusting the dirty heap back into his great-coat pocket. At the next stage a very tidy woman came out, with a rather large bundle, containing fresh linen, she said, for her son, who was ill in the hospital at Timaru. She booked this, and paid her half-crown for its carriage, entreating the drunken wretch to see that it reached her son that night. He wildly promised he should have it in half-an-hour, and we set off as if he meant to keep his word, though we were some forty miles off yet; but he soon changed his mind, and took a hatred to the parcel, saying it would "sink the ship," and finally tried to kick it over the splash-board. I seized it at the risk of losing my balance, and hugged it tight all the way to Timaru, carrying it off to the hotel, where I induced a waiter to take it up to the hospital.

After we had changed horses for the last time, and I was comforting myself by the reflection that the journey was nearly over, we heard shouts and screams from the inside passengers. F— — persuaded Jim with much trouble to pull up, and jumped down to see what was the matter. A strong smell of burning and a good deal of smoke arose from inside the coach, caused by the lunatic having taken off both his boots and lighted a fire in them. It was getting dark and chilly; the other passengers, including the policeman, had dozed off and the madman thought that as his feet were very cold, he would "try and warm them a bit;" so he collected all the newspapers with which his fellow-travellers had been solacing the tedium of their journey, tore, them up into shreds, with the addition of the contents of a poor woman's bundle, and made quite a cheerful blaze out of these materials. It was some time before the terrified women could be induced to get into the coach again; and it was only by Jims asseverations, couched in the strongest language, that if they were not "all aboard" in half a minute, he would drive on and leave them in the middle of the plains, that they were persuaded to clamber in to their places once more.

How thankful I was when we saw the lights of Timaru! I was stunned and bewildered, tired beyond the power of words to describe, and black and blue all over from being jolted about. The road had been an excellent one, all the way level and wide, with telegraph-poles by its side. We shaved these very closely often enough, but certainly, amid all his tipsiness, Jim bore out his predecessors remark. Whenever we came to a little dip in the road, or a sharp turn, as we were nearing Timaru, he would get the horses under control as if by magic, and take us over as safely as the soberest driver could have done; the moment the obstacle was passed, off we were again like a whirlwind!

I was not at all surprised to hear that upsets and accidents were common on the road, and that the horses lasted but a very short time.

We found our host had driven in from his station forty-five miles distant from Timaru, to meet us, and had ordered nice rooms and a good dinner; so the next morning I was quite rested, and ready to laugh over my miseries of the day before. Nothing could be a greater contrast than this day's journeying to yesterday's. A low, comfortable phaeton, and one of the most agreeable companions in the world to drive us, beautiful scenery and a nice luncheon half-way, at which meal F— — ate something like half a hundred cheesecakes! The last part of the road for a dozen miles or so was rather rough; we had to cross a little river, the Waio, every few hundred yards; and a New Zealand river has so much shingle about it! The water can never quite make up its mind where it would like to go, and has half-a-dozen channels ready to choose from, and then in a heavy fresh the chances are it will select and make quite a different course after all.

This is late autumn with us, remember, so the evenings close in early and, are very cold indeed. It was quite dark when we reached the house, and the blazing fires in every room were most welcome. The house is very unlike the conventional station pattern, being built of stone, large, very well arranged, and the perfection of comfort inside. There is no hostess at present; three bachelor brothers do the honours, and, as far as my experience goes, do them most efficiently. Our visit has lasted three weeks already, and we really must bring it to a termination soon. The weather has been beautiful, and we have made many delightful excursions, all on horseback, to neighbouring stations, to a fine bush where we had a picnic, or to some point of view. I can truly say I have enjoyed every moment of the time, indoors as well as out; I was the only lady, and was petted and made much of to my heart's content. There were several other guests, and they were all nice and amusing. One wet day we had, and only one. I must tell you an incident of it, to show you what babies grown-up men can be at the Antipodes. We worked hard all the morning at acrostics, and after my five o'clock tea I went upstairs to a charming little boudoir prepared for me, to rest and read; in a short time I heard something like music and stamping, and, though I was *en peignoir*, I stole softly down to see what was going on; when I opened the door of the general sitting-room a most unusual sight presented itself, — eight bearded men, none of them very young, were dancing a set of quadrilles with the utmost gravity and decorum to the tunes played by a large musical-box, which was going at the most prodigious pace, consequently the dancers were flying through the figures in silence and breathless haste. They could not stop or speak when I came in, and seemed quite surprised at my laughing at them; but you have no idea how ridiculous they looked, especially as their gravity and earnestness were profound.

This is one of the very few stations where pheasants have been introduced, but then, every arrangement has been made for their comfort, and a beautiful house and yard built for their reception on a flat, just beneath the high terrace on which the house stands. More than a hundred young birds were turned out last spring, and there will probably be three times that number at the end of this year. We actually had pheasant twice at dinner; the first, and probably the last time we shall taste game in New Zealand. There is a good deal of thick scrub in the clefts of the home-terrace, and this affords excellent shelter for the young. Their greatest enemies are the hawks, and every variety of trap and cunning device for the destruction of these latter are in use, but as yet without doing much execution among them, they are so wonderfully clever and discerning.

Letter XIX: A Christening gathering. — the fate of Dick.

Broomielaw, June 1867. We reached home quite safely the first week of this month, and I immediately set to work to prepare for the Bishop's visit. We met him at a friend's house one day, just as we were starting homewards, and something led to my telling him about the destitute spiritual condition of my favourite "nest of Cockatoos." With his usual energy, as well as goodness, he immediately volunteered to come up to our little place, hold a service, and christen all the children. We were only too thankful to accept such an offer, as we well knew what an inducement it would be to the people, who would take a great deal of trouble and come from far and near to hear our dear Bishop, who is universally beloved and respected.

For a week beforehand the house smelt all day long like a baker's shop about noon on Sunday, for pies, tarts, cakes, etc., were perpetually being "drawn" from the oven. I borrowed every pie-dish for miles round, and, as on another occasion I have mentioned, plenty of good things which our own resources could not furnish forth came pouring in on all sides with offers to help. F— — and I scoured the country for thirty miles round to invite everybody to come over to us that Sunday; and I think I may truly say everybody came. When I rode over to my "nest" and made the announcement of the Bishop's visit, the people were very much delighted; but a great difficulty arose from the sudden demand for white frocks for all the babies and older children. I rashly promised each child should find a clean white garment awaiting it on its arrival at my house, and took away a memorandum of all the different ages and sizes; the "order" never could have been accomplished without the aid of my sewing-machine. I had a few little frocks by me as patterns, and cut up some very smart white embroidered petticoats which were quite useless to me, to make into little skirts. In spite of all that was going on in the kitchen my maids found time to get these up most beautifully, and by the Saturday night the little bed in the spare room was a heap of snowy small garments, with a name written on paper and pinned to each. The Bishop also arrived quite safely, late that evening, having driven himself up from Christchurch in a little gig.

It is impossible for you to imagine a more beautiful winter's morning than dawned on us that Sunday. A sharp frost over-night only made the air deliciously crisp, for the sun shone so brightly, that by nine o'clock the light film of ice over the ponds had disappeared, and I found the Bishop basking in the verandah when I came out to breakfast, instead of sitting over the blazing wood-fire in the dining-room. We got our meal finished as quickly as possible, and then F— — and Mr. U— — set to work to fill the verandah with forms extemporised out of empty boxes placed at each end, and planks laid across them; every red blanket in the house was pressed into service to cover these rough devices, and the effect at last was quite tidy. By eleven o'clock the drays began to arrive in almost a continual stream; as each came up, its occupants were taken into the kitchen, and given as much as they could eat of cold pies made of either pork or mutton, bread and hot potatoes, and tea. As for teapots, they were discarded, and the tea was made in huge kettles, whilst the milk stood in buckets, into which quart jugs were dipped every five minutes. I took care of all the women and children whilst F— — and Mr. U— — looked after the men, showed them where to put the horses, etc. All this time several gentlemen and two or three ladies had arrived, but there was no one to attend to them, so they all very kindly came out and helped. We insisted on the Bishop keeping quiet in the drawing-room, or he would have worked as hard as any one. I never could have got the children into their white frocks by two o'clock if it had not been for the help of the other ladies; but at last they were all dressed, and the congregation—not much under a hundred people—fed, and arranged in their places. There had been a difficulty about finding sufficient godmothers and godfathers, so F— — and I were sponsors for every child, and each parent wished me to hand the child to the Bishop; but I could not lift up many of the bigger ones, and they roared piteously when I touched their hands. I felt it quite a beautiful and thrilling scene; the sunburnt faces all around, the chubby, pretty little group of white-clad children, every one well fed and comfortably clothed, the dogs lying at their masters feet, the bright winter sunshine and dazzling sky, and our dear Bishops commanding figure and clear, penetrating voice! He gave us a most excellent sermon, short and simple, but so perfectly appropriate; and after the service was over he went about, talking to all the various groups such nice, helpful words.

The truest kindness was now to "speed the parting guest," so each dray load, beginning with those whose homes were the most distant, was collected. They were first taken into the kitchen and given a good meal of hot tea, cake, and bread and butter, for many had four hours' jolting before them; the red blankets were again called into requisition to act as wraps, besides every cloak and shawl I possessed, for the moment the sun sunk, which would be about four o'clock, the cold was sure to become intense. We lived that day in the most scrambling fashion ourselves; there was plenty of cold meat, etc., on the dining-room table, and piles of plates, and whenever any of the party were hungry they went and helped themselves, as my two servants were entirely occupied with looking after the comfort of the congregation; it was such a treat to them to have, even for a few hours, the society of other women. They have only one female neighbour, and she is generally too busy to see much of them; besides which, I think the real reason of the want of intimacy is that Mrs. M— — is a very superior person, and when she comes up I generally like to have a chat with her myself. It does me good to see her bonny Scotch face, and hear the sweet kindly "Scot's tongue;" besides which she is my great instructress in the mysteries of knitting socks and stockings, spinning, making really good butter (not an easy thing, madam), and in all sorts of useful accomplishments; her husband is the head shepherd on the next station. They are both very fond of reading, and it was quite pretty to see the delight they took in the Queen's book about the Highlands.

To return, however, to that Sunday. We were all dreadfully tired by the time the last guest had departed, but we had a delightfully quiet evening, and a long talk with the Bishop about our favourite scheme of the church and school among the Cockatoos, and we may feel certain of his hearty cooperation in any feasible plan for carrying it out. The next morning, much to our regret, the Bishop left us for Christchurch, but he had to hold a Confirmation service there, and could not give us even a few more hours. We were so very fortunate in our weather. The following Sunday was a pouring wet day, and we have had wind and rain almost ever since; it is unusually wet, so I have nothing more to tell you of our doings, which must seem very eccentric to you, by the way, but I assure you I enjoy the gipsy unconventional life immensely.

You must not be critical about a jumble of subjects if I record poor Dick's tragical fate here; it will serve to fill up my letter, and if ever you have mourned for a pet dog you will sympathise with me. I must first explain to you that on a sheep station strange dogs are regarded with a most unfriendly eye by both master and shepherds. There are the proper colleys,—generally each shepherd has two,—but no other dogs are allowed, and I had great trouble to coax F— — to allow me to accept two. One is a beautiful water-spaniel, jet black, Brisk by name, but his character is stainless in the matter of sheep, and though very handsome he is only an amiable idiot, his one amusement being to chase a weka, which he never catches. The other dog was, alas! Dick, a small black-and-tan terrier, very well bred, and full of tricks and play. We never even suspected him of any wickedness, but as it turned out he must have been a hardened offender. A few weeks after he came to us, when the lambing season was at its height, and the low sunny hills near the house were covered with hundreds of the pretty little white creatures, F— —used sometimes to come and ask me where Dick was, and, strange to say, Dick constantly did not answer to my call. An evening or two later, just as we were starting for our walk, Dick appeared in a great hurry from the back of the stable. F— — went up immediately to him, and stooped down to examine his mouth, calling me to see. Oh, horror! it was all covered with blood and wool. I pleaded all sorts of extenuating circumstances, but F— — said, with: judicial sternness, "This cannot be allowed." Dick was more fascinating than usual, never looking at a sheep whilst we were out walking with him, and behaving in the most exemplary manner. F— — watched him all the next day, and at last caught him in the act of killing a new-born lamb a little way from the house; the culprit was brought to me hanging his tail with the most guilty air, and F— — said, "I ought to shoot him, but if you like I will try if a beating can cure him, but it must be a tremendous one." I was obliged to accept this alternative, and retreated where I could not hear Dick's howls under the lash, over the body of his victim. A few hours after I went to the spot, lifted Dick up, and carried him into my room to nurse him; for he could not move, he had been beaten so severely. For two whole days he lay on the soft mat I gave him, only able to lap a little warm milk; on the third morning he tried to get up, and crawled into the verandah; I followed to watch him. Imagine my dismay at seeing him limp to the place where the body of his last victim lay, and deliberately begin tearing it to pieces. I followed him with my little horsewhip and gave him a slight beating. I could not find it in my heart to hit him very hard. I carefully concealed this incident from F— —, and for some days I never let Dick out of my sight for a moment; but early one fine morning a knock came to our bed-room door, and a voice said, "Please, sir, come and see what's the matter with the sheep? there's a large mob of them at the back of the house being driven, like." Oh, my prophetic soul! I felt it was Dick. Whilst F— — was huddling on some clothes I implored him to temper justice with mercy, but never a word did he say, and sternly took his gun in his hand and went out. I buried my head in the pillows, but for all my precautions I heard the report of a shot in the clear morning air, and the echo ringing back from all the hills; five minutes afterwards F— — came in with a little blue collar in his hand, and said briefly, "He has worried more than a dozen lambs this morning alone." What could I say? F— —'s only attempt at consolation was, "he died instantly; I shot him through the head." But for many days afterwards I felt quite lonely and sad without my poor little pet— yet what could have been done? No one would have accepted him as a present, and it flashed on me afterwards that perhaps this vice of his was the reason of Dick's former owner being so anxious to give him to me. I have had two offers of successors to Dick since, but I shall never have another dog on a sheep station, unless I know what Mr. Dickens' little dressmaker calls "its tricks and its manners."

Letter XX: the New Zealand snowstorm of 1867.

Broomielaw, August 1867. I have had my first experience of real hardships since I last wrote to you. Yes, we have all had to endure positive hunger and cold, and, what I found much harder to bear, great anxiety of mind. I think I mentioned that the weather towards the end of July had been unusually disagreeable, but not very cold This wet fortnight had a great deal to do with our sufferings afterwards, for it came exactly at the time we were accustomed to send our dray down to Christchurch for supplies of flour and groceries, and to lay in a good stock of coals for the winter; these latter had been ordered, and were expected every day. Just the last few days of July the weather cleared up, and became like our usual most beautiful winter climate; so, after waiting a day or two, to allow the roads to dry a little, the dray was despatched to town, bearing a long list of orders, and with many injunctions to the driver to return as quickly as possible, for all the stores were at the lowest ebb. I am obliged to tell you these domestic details, in order that you may understand the reason of our privations. I acknowledge, humbly, that it was not good management, but sometimes accidents *will* occur. It was also necessary for F— — to make a journey to Christchurch on business, and as he probably would be detained there for nearly a week, it was arranged that one of the young gentlemen from Rockwood should ride over and escort me back there, to remain during F— —'s absence. I am going to give you all the exact dates, for this snow-storm will be a matter of history, during the present generation at all events: there is no tradition among the Maoris of such a severe one ever having occurred; and what made it more fatal in its financial consequences to every one was, that the lambing season had only just commenced or terminated on most of the runs. Only a few days before he left, F— — had taken me for a ride in the sheltered valleys, that he might see the state of the lambs, and pronounced it most satisfactory; thousands of the pretty little creatures were skipping about by their mothers' side.

I find, by my Diary, July 29th marked, as the beginning of a "sou'-wester." F— — had arranged to start that morning, and as his business was urgent, he did not like to delay his departure, though the day was most unpromising, a steady, fine drizzle, and raw atmosphere; however, we hurried breakfast, and he set off, determining to push on to town as quickly as possible. I never spent such a dismal day in my life: my mind was disturbed by secret anxieties about the possibility of the dray being detained by wet weather, and there was such an extraordinary weight in the air, the dense mist seemed pressing everything down to the ground; however, I drew the sofa to the fire, made up a good blaze (the last I saw for some time), and prepared to pass a lazy day with a book; but I felt so restless and miserable I did not know what was the matter with me. I wandered from window to window, and still the same unusual sight met my eyes; a long procession of ewes and lambs, all travelling steadily down from the hills towards the large flat in front of the house; the bleating was incessant, and added to the intense melancholy of the whole affair. When Mr. U— — came in to dinner; at one o'clock, he agreed with me that it was most unusual weather, and said, that on the other ranges the sheep were drifting before the cold mist and rain just in the same way. Our only anxiety arose from the certainty that the dray would be delayed at least a day, and perhaps two; this was a dreadful idea: for some time past we had been economising our resources to make them last, and we knew that there was absolutely nothing at the home-station, nor at our nearest neighbour's, for they had sent to borrow tea and sugar from us. Just at dusk that evening, two gentlemen rode up, not knowing F— — was from home, and asked if they might remain for the night. I knew them both very well; in fact, one was our cousin T— —, and the other an old friend; so they put up their horses, and housed their dogs (for each had a valuable sheep-dog with him) in a barrel full of clean straw, and we all tried to spend a cheerful evening, but everybody confessed to the same extraordinary depression of spirits that I felt.

When I awoke the next morning, I was not much surprised to see the snow falling thick and fast: no sheep were now visible, there was a great silence, and the oppression in the atmosphere had if possible increased. We had a very poor breakfast,—no porridge, very little mutton (for in expectation of the house being nearly empty, the shepherd had not brought any over the preceding day), and *very* weak tea; coffee and cocoa all finished, and about an ounce of tea in the chest. I don't know how the gentlemen amused themselves that day; I believe they smoked a good deal; I could only afford a small fire in the drawing-room, over which I shivered. The snow continued to fall in dense fine clouds, quite unlike any snow I ever saw before, and towards night I fancied the garden fence was becoming very much dwarfed. Still the consolation was, "Oh, it won't last; New Zealand snow never: does." However, on Wednesday morning things began to look very serious indeed: the snow covered the ground to a depth of four feet in the shallowest places, and still continued to fall steadily; the cows we knew *must* be in the paddock were not to be seen anywhere; the fowl-house and pigstyes which stood towards the weather quarter had entirely disappeared; every scrap of wood (and several logs were lying about at the back) was quite covered up; both the verandahs were impassable; in one the snow was six feet deep, and the only door which could be opened was the back-kitchen door, as that opened inwards; but here the snow was half-way over the roof, so it took a good deal of work with the kitchen-shovel, for no spades could be found, to dig out a passage. Indoors, we were approaching our last mouthful very rapidly, the tea at breakfast was merely coloured hot water, and we had some picnic biscuits with it. For dinner we had the last tin of sardines, the last pot of apricot jam, and a tin of ratifia biscuits a most extraordinary mixture, I admit, but there was nothing else. There were six people to be fed every day, and nothing to feed them with. Thursday's breakfast was a discovered crust of dry bread, very stale, and our dinner that day was rice and salt—the last rice in the store-room. The snow still never ceased falling, and only one window in the house afforded us any light; every box was broken up and used for fuel. The gentlemen used to go all together and cut, or rather dig, a passage through the huge drift in front of the stable, and with much difficulty get some food for the seven starving horses outside, who were keeping a few yards clear by incessantly moving about, the snow making high walls all around them.

It was wonderful to see how completely the whole aspect of the surrounding scenery was changed; the gullies were all filled up, and nearly level with the downs; sharp-pointed cliffs were now round bluffs; there was no vestige of a fence or gate or shrub to be seen, and still the snow came down as if it had only just begun to fall; out of doors the silence was like death, I was told, for I could only peep down the tunnel dug every few hours at the back-kitchen door. My two maids now gave way, and sat clasped in each other's arms all day, crying piteously, and bewailing their fate, asking me whenever I came into the kitchen, which was about every half-hour, for there was no fire elsewhere, "And oh, when do you think we'll be found, mum?" Of course this only referred to the ultimate discovery of our bodies. There was a great search to-day for the cows, but it was useless, the gentlemen sank up to their shoulders in snow. Friday, the same state of things: a little flour had been discovered in a discarded flour-bag, and we had a sort of girdle-cake and water. The only thing remaining in the store-room was some black-lead, and I was considering seriously how that could be cooked, or whether it would be better raw: we were all more than half starved, and quite frozen: very little fire in the kitchen, and none in any other room. Of course, the constant thought was, "Where are the sheep?" Not a sign or sound could be heard. The dogs' kennels were covered several feet deep; so we could not get at them at all. Saturday morning: the first good news I heard was that the cows had been found, and dragged by ropes down to the enclosure the horses had made for them-selves: they were half dead, poor beasts; but after struggling for four hours to and from a haystack two hundred yards off, one end of which was unburied, some oaten hay was procured for them. There was now not a particle of food in the house. The servants remained in their beds, declining to get up, and alleging that they might as well "die warm." In the middle of the day a sort of forlorn-hope was organized by the gentlemen to try to find the fowl-house, but they could not get through the drift: however, they dug a passage to the wash-house, and returned in triumph with about a pound of very rusty bacon they had found hanging up there; this was useless without fuel, so they dug for a little gate leading to the garden, fortunately hit its whereabouts, and soon had it broken up and in the kitchen grate. By dint of taking all the lead out of the tea-chests, shaking it, and collecting every pinch of tea-dust, we got enough to make a teapot of the weakest tea, a cup of which I took to my poor crying maids in their beds, having first put a spoonful of the last bottle of whisky which the house possessed into it, for there was neither, sugar nor milk to be had. At midnight the snow ceased for a few hours, and a hard sharp frost set in; this made our position worse, for they could now make no impression on the snow, and only broke the shovels in trying. I began to think seriously of following the maids example, in order to "die warm." We could do nothing but wait patiently. I went up to a sort of attic where odds and ends were stowed away, in search of something to eat, but could find nothing more tempting than a supply of wax matches. We knew there was a cat under the house, for we heard her mewing; and it was suggested to take up the carpets first, then the boards, and have a hunt for the poor old pussy but we agreed to bear our hunger a little longer, chiefly, I am afraid, because she was known to be both thin and aged.

Towards noon on Sunday the weather suddenly changed, and rain began to come down heavily and steadily; this cheered us all immensely, as it would wash the snow away probably, and so it did to some degree; the highest drifts near the house lessened considerably in a few hours, and the gentlemen, who by this time were desperately hungry, made a final attempt in the direction of the fowl-house, found the roof, tore off some shingles, and returned with a few aged hens, which were mere bundles of feathers after their week's starvation. The servants consented to rise and pluck them, whilst the gentlemen sallied forth once more to the stock-yard, and with great difficulty got off two of the cap or top rails, so we had a splendid though transitory blaze, and some hot stewed fowl; it was more of a soup than anything else, but still we thought it delicious: and then everybody went to bed again, for the house was quite dark still, and the oil and candles were running very low. On Monday morning the snow was washed off the roof a good deal by the deluge of rain which had never ceased to come steadily down, and the windows were cleared a little, just at the top; but we were delighted with the improvement, and some cold weak fowl-soup for breakfast, which we thought excellent. On getting out of doors, the gentlemen reported the creeks to be much swollen and rushing in yellow streams down the sides of the hills over the snow, which was apparently as thick as ever; but it was now easier to get through at the surface, though quite solid for many feet from the ground. A window was scraped clear, through which I could see the desolate landscape out of doors, and some hay was carried with much trouble to the starving cows and horses, but this was a work of almost incredible difficulty. Some more fowls were procured to-day, nearly the last, for a large hole in the roof showed most of them dead of cold and hunger.

We were all in much better spirits on this night, for there were signs of the wind shifting from south to north-west; and, for the first time in our lives I suppose, we were anxiously watching and desiring this change, as it was the only chance of saving the thousands of sheep and lambs we now knew lay buried under the smooth white winding-sheet of snow. Before bedtime we heard the fitful gusts we knew so well, and had never before hailed with such deep joy and thankfulness. Every time I woke the same welcome sound of the roaring warm gale met my ears; and we were prepared for the pleasant sight, on Tuesday morning, of the highest rocks on the hilltops standing out gaunt and bare once more. The wind was blowing the snow off the hills in clouds like spray, and melting it everywhere so rapidly that we began to have a new anxiety, for the creeks were rising fast, and running in wide, angry-looking rivers over the frozen snow on the banks. All immediate apprehension of starvation, however, was removed, for the gentlemen dug a pig out of his stye, where he had been warm and comfortable with plenty of straw, and slaughtered him; and in the loft of the stable was found a bag of Indian meal for fattening poultry, which made excellent cakes of bread. It was very nasty having only ice-cold water to drink at every meal. I especially missed my tea for breakfast; but felt ashamed to grumble, for my disagreeables were very light compared to those of the three gentlemen. From morning to night they were wet through, as the snow of course melted the moment they came indoors. All the first part of the last week they used to work out of doors, trying to get food and fuel, or feeding the horses, in the teeth of a bitter wind, with the snow driving like powdered glass against their smarting hands and faces; and they were as cheery and merry as possible through it all, trying hard to pretend they were neither hungry nor cold, when they must have been both. Going out of doors at this stage of affairs simply meant plunging up to their middle in a slush of half-melted snow which wet them thoroughly in a moment; and they never had dry clothes on again till they changed after dark, when there was no more possibility of outdoor work.

Wednesday morning broke bright and clear for the first time since Sunday week; we actually saw the sun. Although the "norwester" had done so much good for us, and a light wind still blew softly from that quarter, the snow was yet very deep; but I felt in such high spirits that I determined to venture out, and equipped myself in a huge pair of F——'s riding-boots made of kangaroo-skin, well greased with weka-oil to keep the wet out, These I put on over my own thick boots, but my precautions "did nought avail," for the first step I took sank me deep in the snow over the tops of my enormous boots. They filled immediately, and then merely served to keep the snow securely packed round my ankles; however, I struggled bravely on, every now and then sinking up to my shoulders, and having to be hauled out by main force. The first thing done was to dig out the dogs, who assisted the process by vigorously scratching away inside and tunnelling towards us. Poor things! how thin they looked, but they were quite warm; and after indulging in a long drink at the nearest creek, they bounded about, like mad creatures. The only casualties in the kennels were two little puppies, who were lying cuddled up as if they were asleep, but proved to be stiff and cold; and a very old but still valuable collie called "Gipsy." She was enduring such agonies from rheumatism that it was terrible to hear her howls; and after trying to relieve her by rubbing, taking her into the stable-and in fact doing all we could for her—it seemed better and kinder to shoot her two days afterwards.

We now agreed to venture into the paddock and see what had happened to the bathing-place about three hundred yards from the house. I don't think I have told you that the creek had been here dammed up with a sod wall twelve feet high, and a fine deep and broad pond made, which was cleared of weeds and grass, and kept entirely for the gentlemen to have a plunge and swim at daylight of a summer's morning; there had been a wide trench cut about two feet from the top, so as to carry off the water, and hitherto this had answered perfectly. The first thing we had to do was to walk over the high five-barred gate leading into the paddock just the topmost bar was sticking up, but there was not a trace of the little garden-gate or of the fence, which was quite a low one. We were, however, rejoiced to see that on the ridges of the sunny downs there were patches, or rather streaks, of tussocks visible, and they spread in size every moment, for the sun was quite warm, and the "nor'-wester," had done much towards softening the snow. It took us a long time to get down to where the bathing-place *had been*, for the sod wall was quite carried away, and there was now only a heap of ruin, with a muddy torrent pouring through the large gap and washing it still more away. Close to this was a very sunny sheltered down, or rather hill; and as the snow was rapidly melting off its warm sloping sides we agreed to climb it and see if any sheep could be discovered, for up to this time there had been none seen or heard, though we knew several thousands must be on this flat and the adjoining ones.

As soon as we got to the top the first glance showed us a small dusky patch close to the edge of one of the deepest and widest creeks at the bottom of the pad-dock; experienced eyes saw they were sheep, but to me they had not the shape of animals at all, though they were quite near enough to be seen distinctly. I observed the gentlemen exchange looks of alarm, and they said to each other some low words, from which I gathered that they feared the worst. Before we went down to the flat we took a long, careful look round, and made out another patch, dark by comparison with the snow, some two hundred yards lower down the creek, but apparently in the water. On the other side of the little hill the snow seemed to have drifted even more deeply, for the long narrow valley which lay there presented, as far as we could see, one smooth, level snow-field. On the dazzling white surface the least fleck shows, and I can never forget how beautiful some swamp-hens, with their dark blue plumage, short, pert, white tails, and long bright legs, looked, as they searched slowly along the banks of the swollen creek for some traces of their former haunts; but every tuft of tohi-grass lay bent and buried deep beneath its heavy covering. The gentlemen wanted me to go home before they attempted to see the extent of the disaster, which we all felt must be very great; but I found it impossible to do anything but accompany them. I am half glad and half sorry now that I was obstinate; glad because I helped a little at a time when the least help was precious, and sorry because it was really such a horrible sight. Even the first glance showed us that, as soon as we got near the spot we had observed, we were walking on frozen sheep embedded in the snow one over the other; but at all events their misery had been over some time. It was more horrible to see the drowning, or just drowned, huddled-up "mob" (as sheep *en masse* are technically called) which had made the dusky patch we had noticed from the hill.

No one can ever tell how many hundred ewes and lambs had taken refuge under the high terrace which forms the bank of the creek. The snow had soon covered them up, but they probably were quite warm and dry at first. The terrible mischief was caused by the creek rising so rapidly, and, filtering through the snow which it gradually dissolved, drowned them as they stood huddled together. Those nearest the edge of the water of course went first, but we were fortunately in time to save a good many, though the living seemed as nothing compared to the heaps of dead. We did not waste a moment in regrets or idleness; the most experienced of the gentlemen said briefly what was to be done, and took his coat off; the other coats and my little Astrachan jacket were lying by its side in an instant, and we all set to work, sometimes up to our knees in icy water, digging at the bank of snow above us—if you can call it digging when we had nothing but our hands to dig, or rather scratch, with. Oh, how hot we were in five minutes! the sun beating on us, and the reflection from the snow making its rays almost blinding. It was of no use my attempting to rescue the sheep, for I could not move them, even when I had *scrattled* the snow away from one. A sheep, especially with its fleece full of snow, is beyond my small powers: even the lambs I found a tremendous weight, and it must have been very absurd, if an idler had been by, to see me, with a little lamb in my arms, tumbling down at every second step, but still struggling manfully towards the dry oasis where we put each animal as it was dug out. The dear doggies helped us beautifully, working so eagerly and yet so wisely under their master's eye, as patient and gentle with the poor stiffened creatures as if they could feel for them. I was astonished at the vitality of some of the survivors; if they had been very far back and not chilled by the water, they were quite lively. The strongest sheep were put across the stream by the dogs, who were obedient to their master's finger, and not to be induced on any terms to allow the sheep to land a yard to one side of the place on the opposite bank, but just where they were to go. A good many were swept away, but after six hours' work we counted 1,400 rescued ones slowly "trailing" up the low sunny hill I have mentioned, and nibbling at the tussocks as they went. The proportion of lambs was, of course, very small, but the only wonder to me is that there were any alive at all. If I had been able to stop my scratching but for a moment, I would have had what the servants call a "good cry" over one little group I laid bare. Two fine young ewes were standing leaning against each other in a sloping position, like a tent, frozen and immoveable: between them, quite dry, and as lively as a kitten, was a dear little lamb of about a month old belonging to one; the lamb of the other lay curled up at her feet, dead and cold; I really believe they had hit upon this way of keeping the other alive. A more pathetic sight I never beheld.

It is needless to say that we were all most dreadfully exhausted by the time the sun went down, and it began to freeze; nothing but the sheer impossibility of doing anything more in the hardening snow and approaching darkness made us leave off even then, though we had not tasted food all day. The gentlemen took an old ewe, who could not stand, though it was not actually dead, up to the stable and killed it, to give the poor dogs a good meal, and then they had to get some more rails off the stock-yard to cook our own supper of pork and maize.

The next morning was again bright with a warm wind; so the effect of the night's frost soon disappeared, and we were hard at work directly after breakfast. Nothing would induce me to stay at home, but I armed myself with a coal-scoop to dig, and we made our way to the other "mob;" but, alas! there was nothing to do in the way of saving life, for all the sheep were dead. There was a large island formed at a bend in the creek, where the water had swept with such fury round a point as to wash the snow and sheep all away together, till at some little obstacle they began to accumulate in a heap. I counted ninety-two dead ewes in one spot, but I did not stay to count the lambs. We returned to the place where we had been digging the day before, and set the dogs to hunt in the drifts; wherever they began to scratch we shovelled the snow away, and were sure to find sheep either dead or nearly so: however, we liberated a good many more. This sort of work continued till the following Saturday, when F— — returned, having had a most dangerous journey, as the roads are still blocked up in places with snow-drifts; but he was anxious to get back, knowing I must have been going through "hard times." He was terribly shocked at the state of things among the sheep; in Christchurch no definite news had reached them from any quarter: all the coaches were stopped and the telegraph wires broken down by the snow. He arrived about mid-day, and, directly after the meal we still called dinner, started off over the hills to my "nest of Cockatoos," and brought back some of the men with him to help to search for the sheep, and to skin those that were dead as fast as possible. He worked himself all day at the skinning,—a horrible job; but the fleeces were worth something, and soon all the fences, as they began to emerge from the snow, were tapestried with these ghastly skins, and walking became most disagreeable, on account of the evil odours arising every few yards.

We forgot all our personal sufferings in anxiety about the surviving sheep, and when the long-expected dray arrived it seemed a small boon compared to the discovery of a nice little "mob" feeding tranquilly on a sunny spur. It is impossible to estimate our loss until the grand muster at shearing, but we may set it down at half our flock, and *all* our lambs, or at least 90 per cent. of them. Our neighbours are all as busy as we are, so no accurate accounts of their sufferings or losses have reached us; but, to judge by appearances, the distant "back-country" ranges must have felt the storm more severely even than we have; and although the snow did not drift to such a depth on the plains as with us, or lie so long on the ground, they suffered just as much, — for the sheep took shelter under the high river-banks, and the tragedy of the creeks was enacted on a still larger scale; or they drifted along before the first day's gale till they came to a wire fence, and there they were soon covered up, and trampled each other to death. Not only were sheep, but cattle, found dead in hundreds along the fences on the plains. The newspapers give half a million as a rough estimate of the loss among the flocks in this province alone. We have no reliable news from other parts of the island, only vague rumours of the storm having been still more severe in the Province of Otago, which lies to the south, and would be right in its track; the only thing which all are agreed in saying is, that there never has been such a storm before, for the Maories are strong in weather traditions, and though they prophesied this one, it is said they have no legend of anything like it ever having happened.

Letter XXI: Wild cattle hunting in the Kowai Bush.

Mount Torlesse, October 1867. We are staying for a week at a charming little white cottage covered with roses and honeysuckles, nestled under the shadow of this grand mountain, to make some expeditions after wild cattle in the great Kowai Bush. I am afraid that it does not sound a very orderly and feminine occupation, but I enjoy it thoroughly, and have covered myself with glory and honour by my powers of walking all day.

We have already spent three long happy days in the Bush, and although they have not resulted in much slaughter of our big game, still I for one am quite as well pleased as if we had returned laden with as many beeves as used to come in from a border foray. I am not going to inflict an account of each expedition on you; one will serve to give an idea of all, for though there is no monotony in Nature, it may chance that frequent descriptions of her become so, and this I will not risk.

Our ride over here was a sufficiently ridiculous affair, owing to the misbehaviour of the pack-horse, for it was impossible upon this occasion to manage with as little luggage as usual, so we arranged to take a good-sized carpet-bag (a most unheard-of luxury), and on each side of it was to be slung a rifle and a gun, and smaller bags of bullets, shot, and powder-flasks, disposed to the best advantage on the pack-saddle. This was all very well in theory, but when it came to the point, the proper steady old horse who was to bear the pack was not forthcoming! He had taken it into his head to go on a visit to a neighbouring run, so the only available beast was a young chestnut of most uncertain temper. The process of saddling him was a long one, as he objected to each item of his load as soon as it was put on, especially to the guns; but F— — was very patient, and took good care to tie and otherwise fasten everything so that it was impossible for "Master Tucker" (called, I suppose, after the immortal Tommy) to get rid of his load by either kicking or plunging. At last we mounted and rode by a bridle-path among the hills for some twelve miles or so, then across half-a-dozen miles of plain, and finally we forded a river. The hill-track was about as bad as a path could be, with several wide jumps across creeks at the bottom of the numerous deep ravines, or gullies as we call them. F— — rode first—for we could only go in single file—with the detestable Tucker's bridle over his arm; then came the chestnut, with his ears well back, and his eyes all whites, in his efforts to look at his especial aversion, the guns; he kicked all the way down the many hills, and pulled back in the most aggravating manner at each ascent, and when we came to a creek sat down on his tail, refusing to stir. My position was a most trying one; the track was so bad that I would fain have given my mind entirely to my own safety, but instead of this all my attention was centred on Tucker the odious. When we first started I expressed to F— — my fear that Tucker would fairly drag him off his own saddle, and he admitted that it was very likely, adding, "You must flog him." This made me feel that it entirely depended on my efforts whether F— — was to be killed or not, so I provided myself with a small stock-whip in addition to my own little riding-whip, and we set off. From the first yard Tucker objected to go, but there were friendly sticks to urge him on; however, we soon got beyond the reasonable limits of help, and I tried desperately to impress upon Tucker that I was going to be very severe: for this purpose I flourished my stock-whip in a way that drove my own skittish mare nearly frantic, and never touched Tucker, whom F— — was dragging along by main force. At last I gave up the stock-whip, with its unmanageable three yards of lash, and dropped it on the track, to be picked up as we came home. I now tried to hit Tucker with my horse-whip, but he flung his heels up in Helen's face the moment I touched him. I was in perfect despair, very much afraid of a sudden swerve on my mare's part sending us both down the precipice, and in equal dread of seeing F— — pulled off his saddle by Tucker's suddenly planting his fore-feet firmly together: F— — himself, with the expression of a martyr, looking round every now and then to say, "Can't you make him come on?" and I hitting wildly and vainly, feeling all the time that I was worse than useless. At last the bright idea occurred to me to ride nearly alongside of the fiendish Tucker, but a little above him on the hill, so as to be able to strike him fairly without fear of his heels. As far as Tucker was concerned this plan answered perfectly, for he soon found out he had to go; but Helen objected most decidedly to being taken off the comparative safety of the track and made to walk on a slippery, sloping hill, where she could hardly keep her feet; however, we got on much faster this way. Oh, how tired I was of striking Tucker! I don't believe I hurt him much, but I felt quite cruel. When we came to the plain, I begged F— — to let me lead him; so we changed, and there was no holding back on the chestnut's part then; it must have been like the grass and the stones in the fable. I never was more thankful than when that ride was over, though its disagreeables were soon forgotten in the warm welcome we received from our bachelor hosts, and the incessant discussions about the next day's excursion.

We had finished breakfast by seven o'clock the following morning, and were ready to start. Of course the gentlemen were very fussy about their equipments, and hung themselves all over with cartridges and bags of bullets and powder-flasks; then they had to take care that their tobacco-pouches and match-boxes were filled; and lastly, each carried a little flask of brandy or sherry, in case of being lost and having to camp out. I felt quite unconcerned, having only my flask with cold tea in it to see about, and a good walking-stick was easily chosen. My costume may be described as uncompromising, for it had been explained to me that there were no paths but real rough bush walking; so I dispensed with all little feminine adornments even to the dearly-loved chignon, tucked my hair away as if I was going to put on a bathing-cap, and covered it with a Scotch bonnet. The rest of my toilette must have been equally shocking to the eyes of taste, and I have reason to believe the general effect most hideous; but one great comfort was, no one looked at me, they were all too much absorbed in preparations for a great slaughter, and I only came at all upon sufferance; the unexpressed but prevailing dread, I could plainly see, was that I should knock up and become a bore, necessitating an early return home; but I knew better!

An American waggon and some ponies were waiting to take the whole party to the entrance of the bush, about four miles off, and, in spite of having to cross a rough river-bed, which is always a slow process, it did not take us very long to reach our first point. Here we dismounted, just at the edge of the great dense forest, and, with as little delay as possible in fine arrangements, struck into a path or bullock-track, made for about three miles into the bush for the convenience of dragging out the felled trees by ropes or chains attached to bullocks; they are not placed upon a waggon, so you may easily imagine the state the track was in, ploughed up by huge logs of timber dragged on the ground, and by the bullocks' hoofs besides. It was a mere slough with deep holes of mud in it, and we scrambled along its extreme edge, chiefly trusting to the trees on each side, which still lay as they had been felled, the men not considering them good enough to remove. At last we came to a clearing, and I quite despair of making you understand how romantic and lovely this open space in the midst of the tall trees looked that beautiful spring morning. I involuntarily thought of the descriptions in "Paul and Virginia," for the luxuriance of the growth was quite tropical. For about two acres the trees had been nearly all felled, only one or two giants remaining; their stumps were already hidden by clematis and wild creepers of other kinds, or by a sort of fern very like the hart's-tongue, which will only grow on the bark of trees, and its glossy leaves made an exquisite contrast to the rough old root. The "bushmen"—as the men who have bought twenty-acre sections and settled in the bush are called—had scattered English grass-seed all over the rich leafy mould, and the ground was covered with bright green grass, kept short and thick by a few tame goats browsing about. Before us was the steep bank of the river Waimakiriri, and a few yards from its edge stood a picturesque gable-ended little cottage surrounded by a rustic fence, which enclosed a strip of garden gay with common English spring flowers, besides more useful things, potatoes, etc. The river was about two hundred yards broad just here, and though it foamed below us, we could also see it stretching away in the distance almost like a lake, till a great bluff hid it from our eyes. Overhead the trees were alive with flocks of wild pigeons, ka-kas, parroquets, and other birds, chattering and twittering incessantly and as we stood on the steep bank and looked down, I don't think a minute passed without a brace of wild ducks flying past, grey, blue, and Paradise. These latter are the most beautiful plumaged birds I ever saw belonging to the duck tribe, and, when young, are very good eating, quite as delicate as the famous canvas-back. This sight so excited our younger sportsmen that they scrambled down the high precipice, followed by a water-spaniel, and in five minutes had bagged as many brace. We could not give them any more time, for it was past nine o'clock, and we were all eager to start on the serious business of the day; but before we left, the mistress of this charming "bush-hut" insisted on our having some hot coffee and scones and wild honey, a most delicious second breakfast. There was a pretty little girl growing up, and a younger child, both the picture of health; the only drawback seemed to be the mosquitoes; it was not very lonely, for one or two other huts stood in clearings adjoining, and furnished us with three bushmen as guides and assistants. I must say, they were the most picturesque of the party, being all handsome men, dressed in red flannel shirts and leathern knickerbockers and gaiters; they had fine beards, and wore "diggers' hats," a head-dress of American origin—a sort of wide-awake made of plush, capable of being crushed into any shape, and very becoming. All were armed with either rifle or gun, and one carried an axe and a coil of rope; another had a gun such as is seldom seen out of an arsenal; it was an old flint lock, but had been altered to a percussion; its owner was very proud of it, not so much for its intrinsic beauty, though it once had been a costly and splendid weapon and was elaborately inlaid with mother-of-pearl, but because it had belonged to a former Duke of Devonshire. In spite of its claims to consideration on this head as well as its own beauty, we all eyed it with extreme disfavour on account of a peculiarity it possessed of not going off when it was intended to do so, but about five minutes afterwards.

It was suggested to me very politely that I might possibly prefer to remain behind and spend the day in this picturesque spot, but this offer I declined steadily; I think the bushmen objected to my presence more than any one else, as they really meant work, and dreaded having to turn back for a tired "female" (they never spoke of me by any other term). At last all the information was collected about the probable whereabouts of the wild cattle—it was so contradictory, that it must have been difficult to arrange any plan by it,—and we started. A few hundred yards took us past the clearings and into the very heart of the forest. We had left the sun shining brightly overhead; here it was all a "great green gloom." I must describe to you the order in which we marched. First came two of the most experienced "bush-hands," who carried a tomahawk or light axe with which to clear the most cruel of the brambles away, and to notch the trees as a guide to us on our return; and also a compass, for we had to steer for a certain point, the bearings of which we knew—of course the procession was in Indian file: next to these pioneers walked, very cautiously, almost on tiptoe, four of our sportsmen; then I came; and four or five others, less keen or less well armed, brought up the rear. I may here confess that I endured in silence agonies of apprehension for my personal safety all day. It was so dreadful to see a bramble or wild creeper catch in the lock of the rifle before me, and to reflect that, unless its owner was very careful, it might "go off of its own accord," and to know that I was exposed to a similar danger from those behind.

We soon got on the fresh tracks of some cows, and proceeded most cautiously and silently; but it could hardly be called walking, it was alternately pushing through dense undergrowth, crawling beneath, or climbing over, high barricades made by fallen trees. These latter obstacles I found the most difficult, for the bark was so slippery; and once, when with much difficulty I had scrambled up a pile of *debris* at least ten feet high, I incautiously stepped on some rotten wood at the top, and went through it into a sort of deep pit, out of which it was very hard to climb. On comparing notes afterwards, we found, that although we had walked without a moment's cessation for eleven hours during the day, a pedometer only gave twenty-two miles as the distance accomplished. Before we had been in the bush half an hour our faces were terribly scratched and bleeding, and so were the gentlemen's hands; my wrists also suffered, as my gauntlets would not do their duty and lie flat. There were myriads of birds around us, all perfectly tame; many flew from twig to twig, accompanying us with their little pert heads on one side full of curiosity; the only animals we saw were some wild sheep looking very disreputable with their long tails and torn, trailing fleeces of six or seven years' growth. There are supposed to be some hundreds of these in the bush who have strayed into it years ago, when they were lambs, from neighbouring runs. The last man in the silent procession put a match into a dead tree every here and there, to serve as a torch to guide us back in the dark; but this required great judgment for fear of setting the whole forest on fire: the tree required to be full of damp decay, which would only smoulder and not blaze. We intended to steer for a station on the other side of a narrow neck of the Great Bush, ten miles off, as nearly as we could guess, but we made many detours after fresh tracks. Once these hoof-marks led us to the brink of such a pretty creek, exactly like a Scotch burn, wide and noisy, tumbling down from rock to rock, but not very deep. After a whispered consultation, it was determined to follow up this creek to a well-known favourite drinking-place of the cattle, but it was easier walking in the water than on the densely-grown banks, so all the gentlemen stepped in one after another. I hesitated a moment with one's usual cat-like antipathy to wet feet, when a stalwart bushman approached, with rather a victimised air and the remark: "Ye're heavy, nae doot, to carry." I was partly affronted at this prejudgment of the case, and partly determined to show that I was equal to the emergency, for I immediately jumped into the water, frightening myself a good deal by the tremendous splash I made, and meeting reproving glances; and nine heads were shaken violently at me.

Nothing could be more beautiful than the winding banks of this creek, fringed with large ferns in endless variety; it was delightful to see the sun and sky once more overhead, but I cannot say that it was the easiest possible walking, and I soon found out that the cleverest thing to do was to wade a little way behind the shortest gentleman of the party, for when he disappeared in a hole I knew it in time to avoid a similar fate; whereas, as long as I persisted in stalking solemnly after my own tall natural protector, I found that I was always getting into difficulties in unexpectedly deep places. I saw the bushmen whispering together, and examining the rocks in some places, but I found on inquiry that their thoughts were occupied at the moment by other ideas than sport; one of them had been a digger, and was pronouncing an opinion that this creek was very likely to prove a "home of the gold" some day. There is a strong feeling prevalent that gold will be found in great quantities all over the island. At this time of the year the water is very shallow, but the stream evidently comes down with tremendous force in the winter; and they talk of having "found the colour" (of gold) in some places. We proceeded in this way for about three miles, till we reached a beautiful, clear, deep pool, into which the water fell from a height in a little cascade; the banks here were well trodden, and the hoof-prints quite recent; great excitement was caused by hearing a distant lowing, but after much listening, in true Indian fashion, with the ear to the ground, everybody was of a different opinion as to the side from whence the sound proceeded, so we determined to keep on our original course; the compass was once more produced, and we struck into a dense wood of black birch.

Ever since we left the clearing from which the start was made, we had turned our backs on the river, but about three o'clock in the afternoon we came suddenly on it again, and stood on the most beautiful spot I ever saw in my life. We were on the top of a high precipice, densely wooded to the water's edge. Some explorers in bygone days must have camped here, for half-a-dozen trees were felled, and the thick brush-wood had been burnt for a few yards, just enough to let us take in the magnificent view before and around us. Below roared and foamed, among great boulders washed down from the cliff, the Waimakiriri; in the middle of it lay a long narrow strip of white shingle, covered with water in the winter floods, but now shining like snow in the bright sunlight. Beyond this the river flowed as placidly as a lake, in cool green depths, reflecting every leaf of the forest on the high bank or cliff opposite. To our right it stretched away, with round headlands covered with timber running down in soft curves to the water. But on our left was the most perfect composition for a picture in the foreground a great reach of smooth water, except just under the bank we stood on, where the current was strong and rapid; a little sparkling beach, and a vast forest rising up from its narrow border, extending over chain after chain of hills, till they rose to the glacial region, and then the splendid peaks of the snowy range broke the deep blue sky line with their grand outlines.

All this beauty would have been almost too oppressive, it was on such a large scale and the solitude was so intense, if it had not been for the pretty little touch of life and movement afforded by the hut belonging to the station we were bound for. It was only a rough building, made of slabs of wood with cob between; but there was a bit of fence and the corner of a garden and an English grass paddock, which looked about as big as a pocket-handkerchief from where we stood. A horse or two and a couple of cows were tethered near, and we could hear the bark of a dog. A more complete hermitage could not have been desired by Diogenes himself, and for the first time we felt ashamed of invading the recluse in such a formidable body, but ungrudging, open-handed hospitality is so universal in New Zealand that we took courage and began our descent. It really was like walking down the side of a house, and no one could stir a step without at least one arm round a tree. I had no gun to carry, so I clung frantically with both arms to each stem in succession. The steepness of the cliff was the reason we could take in all the beauty of the scene before us, for the forest was as thick as ever; but we could see over the tops of the trees, as the ground dropped sheer down, almost in a straight line from the plateau we had been travelling on all day. As soon as we reached the shingle, on which we had to walk for a few hundred yards, we bethought ourselves of our toilettes; the needle and thread I had brought did good service in making us more presentable. We discovered, however, that our faces were a perfect network of fine scratches, some of which *would* go on bleeding, in spite of cold-water applications. Our boots were nearly dry; and my petticoat, short as it was, proved to be the only damp garment: this was the fault of my first jump into the water.

We put the least scratched and most respectable-looking member of the party in the van, and followed him, amid much barking of dogs, to the low porch; and after hearing a cheery "Come in," answering our modest tap at the door, we trooped in one after the other till the little room was quite full. I never saw such astonishment on any human face as on that of the poor master of the house, who could not stir from his chair by the fire, on account of a bad wound in his leg from an axe. There he sat quite helpless, a moment ago so solitary, arid now finding himself the centre of a large, odd-looking crowd of strangers. He was a middle-aged Scotchman, probably of not a very elevated position in life, and had passed many years in this lonely spot, and yet he showed himself quite equal to the occasion.

After that first uncontrollable look of amazement he did the honours of his poor hut with the utmost courtesy and true good-breeding. His only apology was for being unable to rise from his arm-chair (made out of half a barrel and an old flour-sack by the way); he made us perfectly welcome, took it for granted we were hungry—hunger is a very mild word to express my appetite, for one—called by a loud coo-ee to his man Sandy, to whom he gave orders that the best in the house should be put before us, and then began to inquire by what road we had come, what sport we had, etc., all in the nicest way possible. I never felt more awkward in my life than when I stooped to enter that low doorway, and yet in a minute I was quite at my ease again; but of the whole party I was naturally the one who puzzled him the most. In the first place, I strongly suspect that he had doubts as to my being anything but a boy in a rather long kilt; and when this point was explained, he could not understand what a "female," as he also called me, was doing on a rough hunting expedition. He particularly inquired more than once if I had come of my own free will, and could not understand what pleasure I found in walking so far. Indeed he took it so completely for granted that I must be exhausted, that he immediately began to make plans for F— — and me to stop there all night, offering to give up his "bunk" (some slabs of wood made into a shelf, with a tussock mattress and a blanket), and to sleep himself in his arm-chair.

In the meantime, Sandy was preparing our meal. There was an open hearth with a fine fire, and a big black kettle hanging over it by a hook fastened somewhere up the chimney. As soon as this boiled he went to a chest, or rather locker, and brought a double-handful of tea, which he threw into the kettle; then he took from a cupboard the biggest loaf, of bread I ever saw — a huge thing, which had been baked in a camp-oven — and flapped it down on the table with a bang; next he produced a tin milk-pan, and returned to the cupboard to fetch out by the shank-bone a mutton-ham, which he placed in the milk-dish: a bottle of capital whisky was forthcoming from the same place; a little salt on one newspaper, and brown, or rather *black*, sugar on another, completed the arrangements, and we were politely told by Sandy to "wire in," — digger's phraseology for an invitation to commence, which we did immediately, as soon as we could make an arrangement about the four tin plates and three pannikins. I had one all to myself, but the others managed by twos and threes to each plate. I never had a better luncheon in my life; everything was excellent in its way, and we all possessed what we are told is the best sauce. Large as the supplies were, we left hardly anything, and the more we devoured the more pleased our host seemed. There were no chairs; we sat on logs of trees rudely chopped into something like horse-blocks, but to tired limbs which had known no rest from six hours' walking they seemed delightful. After we had finished our meal, the gentlemen went outside to have half a pipe before setting off again; they dared not smoke whilst we were after the cattle, for fear of their perceiving some unusual smell; and I remained for ten minutes with Mr — —. I found that he was very fond of reading; his few books were all of a good stamp, but he was terribly hard-up for anything which he had not read a hundred times over. I hastily ran over the names of some books of my own, which I offered to lend him for as long a time as he liked: and we made elaborate plans for sending them, of my share in which I took a memorandum. He seemed very grateful at the prospect of having anything new, especially now that he was likely to be laid up for some weeks, and I intend to make every effort to give him this great pleasure as soon as possible.

We exchanged the most hearty farewells when the time of parting came, and our host was most earnest in his entreaties to us to remain; but it was a question of getting out of the bush before dusk, so we could not delay. He sent Sandy to guide us by a rather longer but easier way than climbing up the steep cliff to the place where the little clearing at its edge which I have mentioned had been made; and we dismissed our guide quite happy with contributions from all the tobacco-pouches, for no one had any money with him. We found our way back again by the notches on the trees as long as the light lasted, and when it got too dark to see them easily, the smouldering trunks guided us, and we reached the clearing from which we started in perfect safety. Good Mrs. D— — had a bountiful tea ready; she was much concerned at our having yet some three miles of bad walking before we could reach the hut on the outskirts of the bush, where we had left the trap and the ponies. When we got to this point there was actually another and still more sumptuous meal set out for us, to which, alas! we were unable to do any justice; and then we found our way to the station across the flat, down a steep cutting, and through the river-bed, all in the dark and cold. We had supper as soon as we reached home, tumbling into bed as early as might be afterwards for such a sleep as you Londoners don't know anything about.

I have only described one expedition to you, and that the most unsuccessful, as far as killing anything goes; but my hunting instincts only lead me to the point of reaching the game; when it comes to that, I always try to save its life, and if this can't be done, I retire to a distance and stop my ears; indeed, if very much overexcited, I can't help crying. Consequently, I enjoy myself much more when we don't kill anything; and, on the other occasions, I never could stop and see even the shot fired which was to bring a fine cow or a dear little calf down, but crept away as far as ever I could, and muffled my head in my jacket. The bushmen liked this part of the performance the best, I believe, and acted as butchers very readily, taking home a large joint each to their huts, a welcome change after the eternal pigeons, ka-kas, and wild ducks on which they live.

Letter XXII: The exceeding joy of "burning."

Broomielaw, December 1867. I am quite sorry that the season for setting fire to the long grass, or, as it is technically called, "burning the run," is fairly over at last. It has been later than usual this year, on account of the snow having lain such an unusual time on the ground and kept the grass damp. Generally September is the earliest month in which it begins, and November the latest for it to end; but this year the shady side of "Flagpole" was too moist to take fire until December.

It is useless to think of setting out on a burning expedition unless there is a pretty strong nor'-wester blowing; but it must not be *too* violent, or the flames will fly over the grass, just scorching it instead of making "a clean burn." But when F— — pronounces the wind to be just right, and proposes that we should go to some place where the grass is of two, or, still better, three years' growth, then I am indeed happy. I am obliged to be careful not to have on any inflammable petticoats, even if it is quite a warm day, as they are very dangerous; the wind will shift suddenly perhaps as, I am in the very act of setting a tussock a-blaze, and for half a second I find myself in the middle of the flames. F— — generally gets his beard well singed, and I have nearly lost my eyelashes more than once. We each provide ourselves with a good supply of matches, and on the way we look out for the last year's tall blossom of those horrid prickly bushes called "Spaniards," or a bundle of flax-sticks, or, better than all, the top of a dead and dry Ti-ti palm. As soon as we come to the proper spot, and F— — has ascertained that no sheep are in danger of being made into roast mutton before their time, we begin to light our line of fire, setting one large tussock blazing, lighting our impromptu torches at it, and then starting from this "head-centre," one to the right and the other to the left, dragging the blazing sticks along the grass. It is a very exciting amusement, I assure you, and the effect is beautiful, especially as it grows dusk and the fires are racing up the hills all around us. Every now and then they meet with a puff of wind, which will perhaps strike a great wall of fire rushing up-hill as straight as a line, and divide it into two fiery horns like a crescent; then as the breeze changes again, the tips of flame will gradually approach each other till they meet, and go on again in a solid mass of fire.

If the weather has been very dry for some time and the wind is high, we attempt to burn a great flax swamp, perhaps, in some of the flats. This makes a magnificent bonfire when once it is fairly started, but it is more difficult to light in the first instance, as you have to collect the dead flax-leaves and make a little fire of them under the big green bush in order to coax it to blaze up: but it crackles splendidly; indeed it sounds as if small explosions were going on sometimes. But another disadvantage of burning a swamp is, that there are deep holes every yard or two, into which I always tumble in my excitement, or in getting out of the way of a flax-bush which has flared up just at the wrong moment, and is threatening to set me on fire also. These holes are quite full of water in the winter, but now they contain just enough thin mud to come in over the tops of my boots; so I do not like stepping into one every moment. We start numerous wild ducks and swamp-hens, and perhaps a bittern or two, by these conflagrations. On the whole, I like burning the hill-sides better than the swamp—you get a more satisfactory blaze with less trouble; but I sigh over these degenerate days when the grass is kept short and a third part of a run is burned regularly ever spring, and long for the good old times of a dozen years ago, when the tussocks were six feet high. What a blaze they must have made! The immediate results of our expeditions are vast tracts of perfectly black and barren country, looking desolate and hideous to a degree hardly to be imagined; but after the first spring showers a beautiful tender green tint steals over the bare hill-sides, and by and by they are a mass of delicious young grass, and the especial favourite feeding-place of the ewes and lambs. The day after a good burn thousands of sea-gulls flock to the black ground. Where they spring from I cannot tell, as I never see one at any other time, and their hoarse, incessant cry is the first sign you have of their arrival. They hover over the ground, every moment darting down, for some insect. They cannot find much else but roasted lizards and, grasshoppers, for I have never seen a caterpillar in New Zealand.

In the height of the burning season last month I had Alice S—— to stay with me for two or three weeks, and to my great delight I found our tastes about fires agreed exactly, and we both had the same grievance—that we never were allowed to have half enough of it; so we organized the most delightful expeditions together. We used to have a quiet old station-horse saddled, fasten the luncheon-basket to the pommel with materials for a five o'clock tea, and start off miles away to the back of the run, about three o'clock in the afternoon, having previously bribed the shepherd to tell us where the longest grass was to be found—and this he did very readily, as our going saved him the trouble of a journey thither, and he was not at all anxious for more work than he could help. We used to ride alternately, till we got to a deserted shepherd's hut in such a lovely gully, quite at the far end of the run! Here we tied up dear quiet old Jack to the remnants of the fence, leaving him at liberty to nibble a little grass. We never took off the saddle after the first time, for upon that occasion we found that our united strength was insufficient to girth it on again properly, and we made our appearance at home in the most ignominious fashion—Alice leading Jack, and I walking by his side holding the saddle *on*. Whenever we attempted to buckle the girths, this artful old screw swelled himself out with such a long breath that it was impossible to pull the strap to the proper hole; we could not even get it tight enough to stay steady, without slipping under him at every step. However, this is a digression, and I must take you back to the scene of the fire, and try to make you understand how delightful it was. Alice said that what made it so fascinating to her was a certain sense of its being mischief, and a dim feeling that we might get into a scrape. I don't think I ever stopped to analyse my sensations; fright was the only one I was conscious of, and yet I liked it so much. When after much consultation—in which I always deferred to Alice's superior wisdom and experience—we determined on our line of fire, we set to work vigorously, and the great thing was to see who could make the finest blaze. I used to feel very envious if my fire got into a bare patch, where there were more rocks than tussocks, and languished, whilst Alice's was roaring and rushing up a hill. We always avoided burning where a grove of the pretty Ti-ti palms grew; but sometimes there would be one or two on a hill-side growing by themselves, and then it was most beautiful to see them burn. Even before the flames reached them their long delicate leaves felt the wind of the fire and shivered piteously; then the dry old ones at the base of the stem caught the first spark like tinder, and in a second the whole palm was in a blaze, making a sort of heart to the furnace, as it had so much more substance than the grass. For a moment or two the poor palm would bend and sway, tossing its leaves like fiery plumes in the air, and then it was reduced to a black stump, and the fire swept on up the hill.

The worst of it all was that we never knew when to leave off and come home. We would pause for half an hour and boil our little kettle, and have some tea and cake, and then go on again till quite late, getting well scolded when we reached home at last dead-tired and as black as little chimney-sweeps. One evening F— — was away on a visit of two nights to a distant friend, and Alice and I determined on having splendid burns in his absence; so we made our plans, and everything was favourable, wind and all. We enjoyed ourselves very much, but if Mr. U— — had not come out to look for us at ten o'clock at night, and traced us by our blazing track, we should have had to camp out, for we had no idea where we were, or that we had wandered so many miles from home; nor had we any intention of returning just yet. We were very much ashamed of ourselves upon that occasion, and took care to soften the story considerably before it reached F— —'s ears the next day.

However much I may rejoice at nor'-westers in the early spring as aids to burning the run, I find them a great hindrance to my attempts at a lawn. Twice have we had the ground carefully dug up and prepared; twice has it been sown with the best English seed for the purpose, at some considerable expense; then has come much toil on the part of F— — and Mr. U— — with a heavy garden-roller; and the end of all the trouble has been that a strong nor'-wester has blown both seed and soil away, leaving only the hard un-dug (I wonder whether there is such a word) ground. I could scarcely believe that it really was all "clean gone," as children say, until a month or two after the first venture, when I had been straining my eyes and exercising my imagination all in vain to discover a blade where it ought to have been, but had remarked in one of my walks an irregular patch of nice English grass about half a mile from the house down the flat. I speculated for some time as to how it got there, and at last F— — was roused from his reverie, and said coolly, "Oh, that's your lawn!" When this happens twice, it really becomes very aggravating: there are the croquet things lying idle in the verandah year after year, and, as far as I can see, they are likely to remain unused for ever.

Before I close my letter I must tell you of an adventure I have had with a wild boar, which was really dangerous. F— — and another gentleman were riding with me one afternoon in a very lonely gully at the back of the run, when the dogs (who always accompany us) put up a large, fierce, black, boar out of some thick flax-bushes. Of course the hunting instinct, which all young Englishmen possess, was in full force instantly; and in default of any weapon these two jumped off their horses and picked up, out of the creek close by, the largest and heaviest stones they could lift. I disapproved of the chase under the circumstances, but my timid remonstrances were not even heard. The light riding-whips which each gentleman carried were hastily given to me to hold, and in addition F— — thrust an enormous boulder into my lap, saying, "Now, this is to be my second gun; so keep close to me." Imagine poor me, therefore, with all three whips tucked under my left arm, whilst with my right I tried to keep the big stone on my knee, Miss Helen all the time capering about, as she always does when there is any excitement; and I feeling very unequal to holding her back from joining in the chase too ardently, for she always likes to be first everywhere, which is not at all my "sentiments." The ground was as rough as possible; the creek winding about necessitated a good jump every few yards; and the grass was so long and thick that it was difficult to get through it, or to see any blind creeks or other pitfalls. *Mem.* to burn this next spring.

The pig first turned to bay against a palm-tree, and soon disabled the dogs. You cannot think what a formidable weapon a wild boar's tusk is—the least touch of it cuts like a razor; and they are so swift in their jerks of the head when at bay that in a second they will rip up both dogs and horses: nor are they the least afraid of attacking a man on foot in self-defence; but they seldom or ever strike the first blow. As soon as he had disposed of both the dogs, who lay howling piteously and bleeding on the ground, the boar made at full speed for the spur of a hill close by. The pace was too good to last, especially up-hill; so the gentlemen soon caught him up, and flung their stones at him, but they dared not bring their valuable horses too near for fear of a wound which probably would have lamed them for life; and a heavy, rock or stone is a very unmanageable weapon. I was not therefore at all surprised to see that both shots missed, or only very slightly grazed the pig; but what I confess to being perfectly unprepared for was the boar charging violently down-hill on poor unoffending me, with his head on one side ready for the fatal backward jerk, champing and foaming as he came, with what Mr. Weller would call his "vicked old eye" twinkling with rage. Helen could not realize the situation at all. I tried to turn her, and so get out of the infuriated brute's way; but no, she would press on to meet him and join the other horses at the top of the hill. I had very little control over her, for I was so laden with whips and stones that my hands were useless for the reins. I knew I was in great danger, but at the moment I could only think of my poor pretty mare lamed for life, or even perhaps killed on the spot. I heard one wild shout of warning from above, and I knew the others were galloping to my rescue; but in certainly less than half a minute from the time the boar turned, he had reached me. I slipped the reins over my left elbow, so as to leave my hands free, took my whip in my teeth (I had to drop the others), and lifting the heavy stone with both my hands waited a second till the boar was near enough, leaning well over on the right-hand side of the saddle so as to see what he did. He made for poor Helen's near fore-leg with his head well down, and I could hear his teeth gnashing. Just as he touched her with a prick from his tusk like a stiletto and before he could jerk his head back so as to rip the leg up, I flung my small rock with all the strength I possessed crash on his head: but I could not take a good aim; for the moment Helen felt the stab, she reared straight up on her hind-legs, and as we were going up-hill, I had some trouble to keep myself from slipping off over her tail. However, my rock took some effect, for the pig was so stunned that he dropped on his knees, and before he could recover himself Helen had turned round, still on her hind-legs, as on a pivot, and was plunging and jumping madly down the hill. I could not get back properly into my saddle, nor could I arrange the reins; so I had to stick on anyhow. It was not a case of fine riding at all; I merely clung like a monkey, and F— —, who was coming as fast as he could to me, said he expected to see me on the ground every moment; but, however, I did not come off upon that occasion. Helen was nearly beside herself with terror. I tried to pat her neck and soothe her, but the moment she felt my hand she bounded as if I had struck her, and shivered so much that I thought she must be injured; so the moment F— — could get near her I begged him to look at her fetlock. He led her down to the creek, and washed the place, and examined it carefully, pronouncing, to my great joy, that the tusk had hardly gone in at all—in fact had merely pricked her—and that she was not in the least hurt. I could hardly get the gentlemen to go to the assistance of the poor dogs, one of which was very much hurt. Both F— — and Mr. B— — evidently thought I must have been "kilt intirely," for my situation looked so critical at one moment that they could scarcely be persuaded that neither Helen nor I were in the least hurt. I coaxed F— — that evening to write me a doggerel version of the story for the little boys, which I send you to show them:—

> St. Anne and the pig. You've heard of St. George and the dragon, Or seen them; and what can be finer, In silver or gold on a flagon, With Garrard or Hancock designer? Though we know very little about him (Saints mostly are shrouded in mystery), Britannia can't well do without him, He sets off her shillings and history. And from truth let such tales be defended, Bards at least should bestow them their blessing, As a rich sort of jewel suspended On History when she's done dressing. Some would have her downstairs to the present, In plain facts fresh from critical mangle; But let the nymph make herself pleasant, Here a bracelet, and there with a bangle Such as Bold Robin Hood or Red Riding, Who peasant and prince have delighted, Despite of all social dividing, And the times of their childhood united. Shall New Zealand have never a fable, A rhyme to be sung by the nurses, A romance of a famous Round Table, A "Death of Cock Robin" in verses? Or shall not a scribe be found gracious With pen and with parchment, inditing And setting a-sail down the spacious Deep day stream some suitable writing; Some action, some name so heroic That its sound shall be death to her foemen, And make her militia as stoic As St. George made the Cressy crossbowmen; A royal device for her banners, A reverse for her coinage as splendid, An example of primitive manners When all their simplicity's ended? Here it is, ye isles Antipodean! Leave Britain her great Cappadocian; I'll chant you a latter-day paean, And sing you a saint for devotion, Who on horseback slew also a monster, Though armed with no sharp

lance to stab it, Though no helmet or hauberk ensconced her, But only a hat and a habit. This dame, for her bravery sainted, Set up for all times' adoration, With her picture in poetry painted, Was a lady who lived on a station. Her days — to proceed with the story In duties domestic dividing, But, or else she had never won glory, She now and then went out a-riding. It chanced, with two knights at her stirrup, She swept o'er the grass of the valleys, Heard the brooks run; and heard the birds chirrup, When a boar from the flax-bushes sallies. The cavaliers leaped from their horses; As for weapons, that day neither bore them; So they chose from the swift water-courses Heavy boulders, and held them before them. They gave one as well to the lady: She took it, and placed it undaunted On the pommel, and balanced it steady, While they searched where the animal haunted. A bowshot beyond her were riding The knights, each alert with his missile, But in doubt where the pig went a-hiding, For they had not kept sight of his bristle. When — the tale needs but little enlarging One turned round by chance on his courser; To his horror, the monster was charging At the lady, as if to unhorse her. But his fears for her safety were idle, No heart of a hero beat stouter: She poised the stone, gathered her bridle — A halo, 'tis said, shone about her. With his jaws all extended and horrid, Fierce and foaming, the brute leapt to gore her, When she dropped the rock full on his forehead, And lo! he fell dying before her. There he lay, bristling, tusky, and savage; Such a mouth, as was long ago written; Made Calydon lonely with ravage, By such teeth young Adonis was bitten. Then praise to our new Atalanta, Of the chase and of song spoils be brought her, Whose skill and whose strength did not want a Meleager to finish the slaughter. She is sung, and New Zealand shall take her, Thrice blest to possess such a matron, And give thanks to its first ballad-maker, Who found it a saint for a patron.

Letter XXIII: Concerning a great flood.

Broomielaw, February 1868. Since I last wrote to you we have been nearly washed away, by all the creeks and rivers in the country overflowing their banks! Christchurch particularly was in great danger from the chance of the Waimakiriri returning to its old channel, in which case it would sweep away the town. For several hours half the streets were under water, the people going about in boats, and the Avon was spread out like a lake over its banks for miles. The weather had been unusually sultry for some weeks, and during the last five days the heat had been far greater, even in the hills, than anyone could remember. It is often very hot indeed during the mid-day hours in summer, but a hot night is almost unknown; and, at the elevation we live, there are few evenings in the year when a wood-fire is not acceptable after sunset; as for a blanket at night, that is seldom left off even in the plains, and is certainly necessary in the hills. Every one was anxiously looking for rain, as the grass was getting very dry and the creeks low, and people were beginning to talk of an Australian summer and to prophesy dismal things of a drought. On a Sunday night about eleven o'clock we were all sauntering about out of doors, finding it too hot to remain in the verandah; it was useless to think of going to bed; and F— — and Mr. U— — agreed that some great change in the weather was near. There was a strange stillness and oppression in the air; the very animals had not gone to sleep, but all seemed as restless and wakeful as we were. I remember we discussed the probability of a severe earthquake, for the recent wave at St. Thomas's was in everybody's mind. F— — and I had spent a few days in Christchurch the week before. There was a regular low-fever epidemic there, and, he had returned to the station feeling very unwell; but in this country illness is so rare that one almost forgets that such a thing exists, and we both attributed his seediness to the extraordinary heat.

When we were out of doors that Sunday evening, we noticed immense banks and masses of clouds, but they were not in the quarter from whence our usual heavy rain comes; and besides, in New Zealand clouds are more frequently a sign of high wind than of rain. However, about midnight F— — felt so ill that he went in to bed, and we had scarcely got under shelter when, after a very few premonitory drops, the rain came down literally in sheets. Almost from the first F— — spoke of the peculiar and different sound on the roof, but as he had a great deal of fever that night, I was too anxious to notice anything but the welcome fact that the rain had come at last, and too glad to hear it to be critical about the sound it made in falling. I came out to breakfast alone, leaving F— — still ill, but the fever going off. The atmosphere was much lightened, but the rain seemed like a solid wall of water falling fast and furiously; the noise on the wooden roof was so great that we had to shout to each other to make ourselves heard; and when I looked out I was astonished to see the dimensions to which the ponds had. swollen. Down all the hill-sides new creeks and waterfalls had sprung into existence during the night. As soon as I had taken F— — his tea and settled down comfortably to breakfast, I noticed that instead of Mr. U— — looking the picture of bright good-humour, he wore a troubled and anxious countenance. I immediately inquired if he had been out of doors that morning? Yes, he had been to look at the horses in the stable. Well, I did not feel much interest in them, for they were big enough to take care of themselves: so I proceeded to ask if he had chanced to see anything of my fifty young ducks or my numerous broods of chickens. Upon this question Mr. U— — looked still more unhappy and tried to turn the conversation, but my suspicions were aroused and I persisted; so at last he broke to me, with much precaution, that I was absolutely without a duckling or a chicken in the world! They had been drowned in the night, and nothing was to be seen but countless draggled little corpses, what Mr. Mantilini called "moist unpleasant bodies," floating on the pond or whirling in the eddies of the creek. That was not even the worst. Every one of my sitting hens was drowned also, their nests washed away; so were the half-dozen beautiful ducks, with some twelve or fourteen eggs under each. I felt angry with the ducks, and thought they might have at any rate saved their own lives; but nothing could alter the melancholy returns of the missing and dead. My poultry-yard was, for all practical purposes, annihilated, just as it was at its greatest perfection and the pride and joy of my heart. All that day the rain descended steadily in torrents; there was not the slightest break or variation in the downpour: it was as heavy as that of the Jamaica *seasons* of May and October. F— —'s fever left him at the end of twelve hours, and he got up and came into the drawing-room; his first glance out of the window, which commanded a view of the flat for two or three miles, showed him how much the waters had risen since midnight; and he said that in all the years he had known those particular creeks he had never seen them so high: still I thought nothing of it. There was no cessation in the rain for exactly twenty-four hours; but at midnight on Monday, just as poor F— — was getting another attack of fever, it changed into heavy, broken showers, with little pauses of fine drizzle between, and by morning it showed signs of clearing, but continued at intervals till midday. The effect was extraordinary, considering the comparatively short time the real downpour had lasted. The whole flat was under water, the creeks were flooded beyond their banks for half a mile or so on each side, and the river Selwyn, which ran under some hills, bounding our view, was spread out, forming an enormous lake. A very conspicuous object on these opposite hills, which are between three and four miles distant, was a bold cliff known by the name of the "White Rocks," and serving as a landmark to all the countryside: we could hardly believe our eyes when we missed the most prominent of these and could see only a great bare rent in the mountain. The house was quite surrounded by water and stood on a small island; it was impossible even to wade for more than a few yards beyond the dry ground, for the water became quite deep and the current was running fast. F— —'s fever lasted its twelve hours; but I began to be fidgety at the state of prostration it left him in, and when Tuesday night brought a third and sharper attack, I determined to make him go to town and see a doctor during his next interval of freedom from it.

Wednesday morning was bright and sunny, but the waters had not much diminished: however, we knew every hour must lessen them, and I only waited for F— —'s paroxysm of fever to subside about mid-day to send him off to Christchurch. I had exhausted my simple remedies, consisting of a spoonful of sweet spirits of nitre and a little weak brandy and water and did not think it right to let things go on in this way without advice: he was so weak he could hardly mount his horse; indeed he had to be fairly lifted on the old quiet station hack I have before mentioned with such deep affection, dear old Jack. It was impossible for him to go alone; so the ever-kind and considerate Mr. U— — offered to accompany him. This was the greatest comfort to me, though I and my two maids would be left all alone during their absence: however, that was much better than poor F— — going by himself in his weak state. Six hours of sunshine had greatly abated the floods, and as far as we could see the water was quite shallow now where it had overflowed. I saw them set off therefore with a good hope of their accomplishing the journey safely. Judge of my astonishment and horror when, on going to see what the dogs were barking at, about two hours later, I beheld F— — and Mr. U— — at the garden gate, dripping wet up to their shoulders, but laughing very much. Of course I immediately thought of F— —'s fever, and made him come in and change; and have some hot tea directly; but he would not go to bed as I suggested, declaring that the shock of his unexpected cold bath, and the excitement of a swim for his life, had done him all the good in the world; and I may tell you at once; that it had completely cured him: he ate well that evening, slept well, and had no return of his fever, regaining his strength completely in a few days. So much for kill-or-cure remedies!

It seems that as soon as they neared the first creek, with very high banks, about a mile from the house, the water came up to the horses' fetlocks, then to their knees, but still it was impossible to tell exactly where the creek began, or rather, where its bank ended; they went very cautiously, steering as well as they could for where they imagined the cutting in the steep bank to be; but I suppose they did not hit it off exactly, for suddenly they went plump into deep water and found themselves whirling along like straws down a tremendous current. Jack was, however, quite equal to the occasion; he never allows himself to be flurried or put out by anything, and has, I imagine, been in nearly every difficulty incident to New Zealand travelling. Instead, therefore, of losing his head as Helen did (Mr. U— — was riding her), and striking out wildly with her forelegs to the great danger of the other horse, Jack took it all as a matter of course, and set himself to swim steadily down the stream, avoiding the eddies as much as possible: he knew every yard of the bank, and did not therefore waste his strength by trying to land in impossible places, but kept a watchful eye for the easiest spot. F— — knew the old horse so well that he let him have his head and guide himself, only trying to avoid Helen's forelegs, which were often unpleasantly near; his only fear was lest they should have to go so far before a landing was possible that poor old Jack's strength might not hold out, for there is nothing so fatiguing to a horse as swimming in a strong current with a rider on his back, especially a heavy man. They were swept down for a long distance, though it was impossible to guess exactly how far they had gone, and F— — was getting very uneasy about a certain wire fence which had been carried across the creek; they were rapidly approaching it, and the danger was that the horses might suddenly find themselves entangled in it, in which case the riders would very likely have been drowned. F— — called to Mr. U— —to get his feet free from the stirrups and loosened his own; but he told me he was afraid lest Mr. U— — should not hear him above the roaring of the water, and so perhaps be dragged under water when the fence was reached. However, Jack, knew all about it, and was not going to be drowned ignominiously in a creek which would not have wet his hoofs to cross three days before. A few yards from the fence he made one rush and a bound towards what seemed only a clump of Tohi bushes, but they broke the force of the current and gave him the chance he wanted, and he struggled up the high crumbling bank more like a cat than a steady old screw. Helen would not be left behind, and, with a good spur from Mr. U— —, she followed Jack's example, and they stood dripping and shivering in shallow water. Both the horses were so *done* that F— — and Mr. U— — had to jump off instantly and loose the girths, turning them with their nostrils to the wind. It was a very narrow escape, and the disagreeable part of it was that they had scrambled out on the wrong side of the creek and had to recross it to get home: however, they rode on to the next stream, which looked so much more swollen and angry, that they gave up the idea of going on to Christchurch that night, especially as they were wet through to their chins, for both horses swam very low in the water, with only their heads to be seen above it.

The next thing to be considered was how to get back to the house. It never would do to risk taking the horses into danger again when they were so exhausted; so they rode round by the homestead, crossed the creek higher up, where it was much wider but comparatively shallow (if anything could be called shallow just now), and came home over the hills. Good old Jack had an extra feed of oats that evening, a reward to which he is by no means insensible; and indeed it probably is the only one he cares for.

The Fates had determined, apparently, that I also should come in for my share of watery adventures, for we had an engagement of rather long standing to ride across the hills, and visit a friend's station about twelve miles distant, and the day we had promised to go was rather more than a week after F— —'s attempted journey. In the meantime, the waters had of course gone down considerably, and there was quite an excitement in riding and walking about our own run, and seeing the changes the flood had made, and the mischief it had done to the fencing;—this was in process of being repaired. We lost very few sheep; they were all up at the tops of the high hills, their favourite summer pasture.

I think I have told you that between us and Christchurch there is but one river, a most peaceable and orderly stream, a perfect pattern to the eccentric New Zealand rivers, which are so changeable and restless. Upon this occasion, however, the Selwyn behaved quite as badly as any of its fellows; it was not only flooded for miles, carrying away quantities of fencing near its banks, and drowning confiding sheep suddenly, but at one spot about four miles from us, just under the White Rocks, it came down suddenly, like what Miss Ingelow calls "a mighty eygre," and deserted its old timeworn bed for two new ones: and the worst of the story is that it has taken a fancy to our road, swept away a good deal of it, breaking a course for itself in quite a different place; so now, instead of one nice, wide, generally shallow river to cross, about which there never has been an evil report, we have two horrid mountain torrents of which we know nothing: no one has been in yet to try their depth, or to find out the best place at which to ford them, and it unfortunately happened that F— — and I were the pioneers. When we came to the first new channel, F— —with much care picked out what seemed the best place, and though it was a most disagreeable bit of water to go through, still we managed it all right; but when we came to the next curve, it was far worse. Here the river took a sharp turn, and came tearing round a corner, the colour and consistency of pea-soup, and making such a noise we could hardly hear ourselves speak standing close together on the bank; once in the stream, of course it would be hopeless to try to catch a word. I am ashamed to say that my fixed idea was to turn back, and this I proposed without hesitation; but F— — has the greatest dislike to retracing his steps, and is disagreeably like Excelsior in this respect; so he merely looked astonished at my want of spirit, and proceeded very calmly to give me my directions, and the more he impressed the necessity of coolness and caution upon me, the more I quaked. He was to go over first, alone; I was to follow, having first tucked my habit well up under my arm, and taken care that I was quite free so as not to be entangled in any way *if* Helen should be swept away, or if a boulder should come down with the stream, and knock her feet from under her: I was not to be at all frightened (!), and I was to keep my eyes fixed on him, and guide Helen's head exactly by the motion of his hand. He plunged into the water as soon as he had issued these encouraging directions; I saw him floundering in and out of several deep holes, and presently he got safe to land, dripping wet; then he dismounted, tied Leo to a flax bush, and took off his coat and big riding-boots,—I thought, very naturally to dry them, but I should have been still more alarmed, if possible, had I known that this was to prepare to be ready to swim to my help in case of danger. As it was, my only hope was that Helen might not like the look of the angry flood, and would refuse to go in;—how I should have blessed her for such obstinacy!—but no, she was eager to rejoin her stable companion, and plunged in without hesitation. I found it much worse even than I dreaded; the water felt so resistless, as if it *must* sweep me right out of the saddle; I should like to have clutched Helen's mane or anything to have kept me on, but both hands were wanted to hold the reins quite low down, one on each side of her withers, so as to guide her exactly according to F— —'s pilot-hand on the opposite bank: steering implicitly by this I escaped the holes and rocks which he had come against, and got over safely, but trembling, and with chattering teeth. F— —said, quite disdainfully, "You don't mean to say you're really frightened?" So then I scolded him, rather incoherently, and demanded to be praised for coming at all! I wrung my habit out as well as I could, F— — poured the water out of his boots, and we proceeded, first over a plain, and then to climb a high steep hill. I wonder if you have any idea how disagreeable and dangerous it is to go zigzag up the side of a mountain after such rain as we have had. The soil was just like soap, nothing for the horses' hoofs to take hold of, not a pebble or a tuft of grass; all had been washed away, and only the slippery clay remained. As usual, F— — went first and I followed, taking care not to keep below him, lest he and Leo should come "slithering" (that is the only word for it) down upon me; but, alas, it was Helen and I who slithered! Poor dear, all her legs seemed to fly from under her at once, and she came down on her side and on my legs. I felt the leaping-crutch snap, and found my left shoulder against the ground; I let go the reins, and thought we had better part company, but found I could not move for her weight; *she* struggled to get up, and we both slipped down, down—down: there was no reason why we should not have gone on to the bottom of the hill, when a friendly tussock afforded her an instant's resting-place for her hind hoofs, and she scrambled to her feet like a cat. I found myself still on her back; so I picked up my reins and tried to pretend that I had never thought of getting off. F— — dared not stir from his "bad eminence;" so Helen and I wended our slippery way up to him, and in answer to his horrified "Where is your habit?" I found I was torn to ribbons; in fact, my skirt was little more than a kilt, and a very short one too! What was to be done? We were only three or four miles from our destination, so we pushed on, and at the last I lingered behind, and made F— — go first and borrow a cloak or shawl. You would have laughed if you had heard my pathetic adjurations to him to be sure to bring it by himself. I was so afraid that some one else would politely insist on accompanying him. But it was all right, though even with this assistance it was very difficult to arrange matters so as to be tolerably respectable. My hostess was shocked at my tattered, wet plight, and dried me, and dressed me up till I was quite smart, and then we had a very pleasant day, and, best of all, came home by a different road, so as to avoid the slippery descent and the rivers in the dark; but I still mourn for my habit!-it was my last. Three have disappeared, owing to unfortunate accidents, this year, and now I am reduced to what can be contrived out of a linsey dress.

Letter XXIV: My only fall from horseback.

Broomielaw, June 1868. The autumn has passed away so quickly that I can hardly believe the winter has reached us so soon—the last winter we shall spend in New Zealand. I should like to have been able to boast, on my return to England, that in three years' constant riding, on all sorts of horses, good, bad, and indifferent, and over abominable roads, I had escaped a fall; but not only have I had a very severe one, but it was from my own favourite Helen, which is very trying to reflect upon. However, it was not in the least her fault, or mine either; so she and I are still perfectly good friends.

We had been spending two days up at Lake Coleridge, as a sort of farewell visit, and on our way down again to Rockwood, a distance of about twenty miles, we stopped to lunch, by invitation, at a station midway. There was so much to be seen at this place that we loitered much longer than was prudent in the short days, and by the time we had thoroughly inspected a beautiful new wool-shed with all the latest improvements (from which F— — could hardly tear himself away), the fish-ponds elaborately arranged for the reception of the young trout expected from Tasmania and the charming garden well sheltered by a grove of large wattle-trees, it was growing dusk, and we prepared to push on as fast as possible; for nothing is more disagreeable than being caught in the dark on a New Zealand track, with its creeks and swamps and wire fences: the last are the most dangerous obstacles, if you get off the track, or if the gate through the fence has been placed for convenience a few yards on one side of it; the horses cannot see the slender wires in the dark, and so fall over them, injuring themselves and their riders most seriously sometimes. Having still about eight miles to go, we were galloping gaily over a wide open plain, our only anxiety arising from the fast failing daylight; but the horses were still quite fresh, and, as the French idiom would have it, devoured the ground at a fine pace; when, in an instant, the ground appeared to rise up to meet me, and I found myself dragged along on the extreme point of my right shoulder, still grasping both reins and whip. I was almost under the feet of the other horse, and I saw Helen's heels describing frantic circles in the air. F— — shouted to me to let go, which it had never occurred to me to do previously. I did so, and jumped up instantly, feeling quite unhurt, and rather relieved to find that a fall was not so dreadful after all. I then saw the cause of the accident: the handle of a little travelling-bag which had been hung over the pommel of my saddle had slipped over the slight projection, and as it was still further secured by a strap through the girth, it was dangling under poor Helen, whose frantic bounds and leaps only increased the liveliness of her tormentor. I never saw such bucks and jumps high into the air as she performed receiving a severe blow from the bag at each; it was impossible to help laughing, though I did not see how it was all to end. She would not allow F— — to approach her, and was perfectly mad with terror. At last the girths gave way, and the saddle came off, with the bag still fastened to it; the moment she found herself free, she trotted up to me in the most engaging manner, and stood rubbing her nose against my arm, though she was still trembling all over, and covered with foam.

By this time I had made the discovery that I could not raise my right arm; but still a careful investigation did not tell me it was broken, for it gave me no pain to touch anywhere, except a very little just on the point of the shoulder. F— — now went to pick up the saddle and the reins; it was difficult to find these latter in the fast gathering darkness and I held his horse for him. To my horror I found after standing for a moment or two, that I was going to faint; I could not utter a word; I knew that if my fast-relaxing fingers let go their hold of the bridle the horse would set off towards home at a gallop, Helen would assuredly follow him, and we should be left eight miles from the nearest shelter to find our way to it, with a deep creek to cross. F— — was fifty yards off, with his back to me, searching for some indispensable buckle; so there was no help to be got from him at the moment. I exerted every atom of my remaining strength to slip the bridle over my left arm, which I pressed against my waist; then I sat down as quietly as I could, not to alarm the horse, bent forward so as to keep my left arm under me lest the bridle should slip off, and fainted away in great peace and comfort. The cold was becoming so intense that it soon revived me, and F— —, suspecting something was wrong, came to relieve me of the care of the horse, and contrived to get the girths repaired with the ever-ready flax, and the bag secured in a very short time. But when it came to mounting again, that was not so easy: every time I tried to spring, something jarred horribly in the socket where the arm fits into the shoulder, and the pain was so great that I had to lie down on the ground. It was now nearly seven o'clock, quite dark, and freezing hard; we were most anxious to get on, and yet what was to be done? I could not mount, apparently, and there was no stone or bank to stand on and get up by for an immense way. At last F— — put me up by sheer strength. I found myself so deadly sick and faint when I was fairly in the saddle that it was some time before I could allow Helen to move; and never shall I forget the torture of her first step, for my shoulder was now stiffening in a most unpleasant way. F— — said it would be easier to canter; so we set off at full speed, and the cold air against my face kept me from fainting as we went along, though I fully expected to fall off every moment; if Helen had shied, or stumbled, or even capered a little, I should have been on the ground again. In my torture and despair, I proposed to be left behind, and for F— — to ride on and get help; but he would not hear of this, declaring that I should die of cold before he could get back with a cart, and that it was very doubtful if he should find me again on the vast plain, with nothing to guide him, and in the midnight darkness. Whenever we came to a little creek which we were obliged to jump, Helen's safe arrival on the opposite bank was announced by a loud yell from me, caused by agony hardly to be described. The cold appeared to get *into* the broken joint, and make it so much worse.

At last we reached Rockwood, and never was its friendly shelter more welcome. Everything that could be thought of was done to alleviate my sufferings; but I resembled Punch with his head on one side, for I had a well-defined and gigantic hump on my back, and my shoulder was swollen up to my ear. The habit-body was unpicked, as it was impossible to get it off any other way. Of course, the night was one of great agony; but I thought often, as I paced the room, how much better it was to have a blazing fire to cheer me up, and some delicious tea to put my lips to "when so dispoged" (like the immortal Mrs. Gamp), than to be lying on the open plain in a hard frost, wondering when F— — and his cart would arrive.

The next day we returned home, much against our host's wish; and I walked all the way, some six miles of mountain road, for I could not bear the idea of riding. F— — led the horses, and we arrived quite safely. His first idea was to take me down to a doctor, but the motion of driving was greater agony than riding, as the road was rough; so after the first mile, I entreated to be taken back, and we turned the horses' heads towards home again; and when we reached it, I got out all my little books on surgery, medicine, etc., and from them made out how to set my shoulder in some sort of fashion, with F— —'s help. Of course it is still useless to me, but I think it is mending itself; and after a week I could do everything with my left hand, even to writing, after a fashion. The only thing I could *not* do was to arrange my hair, or even to brush it; and though F— — was "willing," he was so exceedingly awkward, that at last, after going through great anguish and having it pulled out by handfuls, I got him to cut it off, and it is now cropped like a small boy's. He cuts up my dinner, etc. for me; but it is a very trying process, and I don't wonder at children often leaving the nasty cold mess half eaten. I shall be very glad to be able to use my own knife again.

Letter XXV: How We lost our horses and had to walk home.

Broomielaw, November 1868. This will actually be my last letter from the Malvern Hills; and, in spite of the joy I feel at the hope of seeing all my beloved ones in England, I am *so* sorry to leave my dear little happy valley. We have done nothing but pay farewell visits lately; and I turn for a final look at each station or cottage as we ride away with a great tightness at my heart, and moisture in my eyes, to think I shall never see them again. You must not be jealous at the lingering regrets I feel, for unless you had been with me here you can never understand how kind and friendly all our neighbours, high and low, have been to us from the very first, or how dearly I have grown to love them. I don't at all know how I am to say good-bye to my dear Mrs. M— —, the shepherd's wife I told you of. I believe she will miss me more than any one; and I cannot bear to think of her left to pass her days without the help of books and papers, which I was always so glad to lend her. I often walk down the valley to take tea with her of an afternoon and to say good-bye, but I have not said it yet. I wish you could see her parlour as I saw it yesterday afternoon—her books in a bookcase of her husband's manufacture, very nice and pretty; her spinning-wheel in the corner; the large "beau-pot" of flowers in the window; and such a tea on the table!—cream like clots of gold, scones, oat-cakes, all sorts of delicacies! She herself is quite charming—one of Nature's ladies. I have given her, as a parting gift, a couple of Scotch views framed; and they hang on the wall as a memento of places equally dear to both of us.

It is a sorrow to me to leave the horses and dogs and my pet calves and poultry; even the trees and creepers I go round to look at, with the melancholy feeling of other owners not loving them so much as I have done. However, I must not make my last letter too dismal, or you will feel that I am not glad enough to return to you all. My only apology is, I have been so *very* happy here.

Now for our latest adventure, as absurd as any, in its way. Have I ever told you that our post-office is ten miles off, with an atrocious road between us and it? I know you will throw down this letter and feel rather disgusted with me for being sorry to leave such a place, but we don't mind trifles here. Lately, since our own establishment has been broken up, we have been living in great discomfort; and among other things we generally, if not always, have to go for our own letters twice a week. Upon this occasion F— — and I had ridden together up the gorge of the Selwyn rather late in the afternoon, to avoid the extreme heat of the day. When we reached the shepherd's hut I have before mentioned, and which is now deserted, I proposed to F— — to go on over the hills alone and leave me there, as I was very hot and tired, and he could travel much quicker without me—for I am ashamed to say that I still object to riding fast up and down slippery hills. I cannot get rid of the idea that I shall break my neck if I attempt it, whereas F— — goes on over the worst road just as if it was perfectly level. Excuse this digression, for it is a relief to me to be a little spiteful about his pace whenever I have an opportunity, and it will probably be my last chance of expressing my entire disapproval of it.

Helen was tied up to a post, and F— —, after helping me to dismount, set off at a canter over the adjoining swamp on his way to cross the chain of hills between the river and the flat where the great coach-road to the West Coast runs. I had brought the ingredients for my five o'clock tea (without which I am always a lost and miserable creature), and I amused myself, during my solitude, by picking up dry bits of scrub for my fire; but I had to go down the river-bank for some driftwood to make the old kettle, belonging to the hut, boil. I could not help wondering how any human being could endure such solitude for years, as the occupant of a hut like this is necessarily condemned to. In itself it was as snug and comfortable as possible, with a little paddock for the shepherd's horse, an acre or so of garden, now overgrown with self-sown potatoes, peas, strawberry, raspberry, and gooseberry plants, the little thatched fowl-house near, and the dog-kennels; all giving it a thoroughly home-like look. The hoarse roar of the river over its rocky bed was the only sound; now and then a flock of wild ducks would come flying down to their roosting-place or nests among the Tohi grass; and as the evening closed in the melancholy cry of the bittern and the weka's loud call broke the stillness, but only to make it appear more profound. On each side of the ravine in which the hut stands rise lofty hills so steeply from the water's edge that in places we can find no footing for our horses, and have to ride in the river. At this time of the year the sheep are all upon the hills; so you do not hear even a bleat: but in winter, they come down to the sunny, sheltered flats.

It appeared to me as if I was alone there for hours, though it really was less than one hour, when F— — returned with a large bundle of letters and papers tied to his saddle-bow. Tea was quite ready now; so he tied up his horse next Helen, and we had tea and looked at our letters. One of the first I opened told me that some friends from Christchurch, whom I expected to pay us a visit soon, were on their way up that very day, and in fact might be expected to arrive just about that hour. I was filled with blank dismay, for not only did the party consist of three grown-up people—nay, four—but three little children. I had made elaborate plans in my head as to how and where they should all be stowed away for a fortnight, but had naturally deferred till the last moment to carry out my arrangements, for they entailed giving up our own bedroom, and "camping" in the dining-room, besides wonderful substitutes of big packing-cases for cribs, etc. etc. But, alas! here we were eight miles from home and nothing done, not even any extra food ordered or prepared. The obvious thing was to mount our horses and return as fast as ever we could, and we hastened out of the hut to the spot where we had left them both securely tied to the only available post, through which unfortunately five wires ran, as it was one of the "standards" of a fence which extended for miles. Just as we came out of the hut in a great bustle, our evil destiny induced F— —'s horse to rub its nose against the top wire of the fence; and in this process it caught the bar of its snaffle-bit, and immediately pulled back: this made all the wires jingle. Helen instantly took alarm, and pulled back too: fresh and increased vibration, extending up the hill-side and echoing back an appalling sound, was the result of this movement. In an instant there were both the horses pulling with all their force against the fence, terrified to death; and no wonder, for the more they pulled the more the wires jingled. F— — did all he could to soothe them with blandishments. I tried to coax Helen, but the nearer we drew the more frantically they backed and plunged, and the more the noise increased—till it was a case of "one struggle more and I am free;" and leaving their bridles still fastened to the fatal fence by the reins, we had the satisfaction of seeing both our horses careering wildly about—first celebrating their escape from danger by joyous and frantic bounds and kicks, and then setting off down the gorge of the river as hard as they could go. I fairly sat down and whimpered a little, not only at the thought of our eight miles' walk over shingle with a deep river to be crossed nine times, but at the idea of my poor little guests arriving to find no supper, no beds, "no nothing."

F— — tried to cheer me up, and said the only thing was to get home as quick as possible; but he did not expect to find that our friends had arrived, for it had been very hazy over the plains all day, and probably had rained hard in Christchurch; so he thought they would not have started on their journey at all. But I refused to accept any comfort from this idea, and bemoaned myself, entirely on their account, incessantly. When we came to the first crossing, F— — picked me up and carried me over dry-shod, and this he did at all the fords; but in one we very nearly came to grief, for I was tilted like a sack over his shoulder, and when we were quite in the middle, and the water was very deep, up to his waist, he kept hoisting my feet higher and higher, quite forgetting that there was plenty more of me on the other side of his shoulder; so it ended in my arms getting very wet, which he did not seem to think mattered at all so long as my feet were dry; whereas I rather preferred having my feet than my head plunged into a surging, deafening yellow current. At the entrance of the gorge is a large stockyard, and near to it, at least a mile or two off, a large mob of horses is generally to be found feeding. We heard great neighing and galloping about amongst them as we came out of the gorge; it was much too dark to distinguish anything, but we guessed that our horses had joined these, and the sounds we heard were probably those of welcome. But the whole mob set off the moment we came near, and crossed the river again, entailing a tenth wetting upon poor F— —. I was posted at the entrance of the gorge, with instructions to shout and otherwise keep them from going up by the route we had just come; but it was more than an hour before F— —could get round the wary brutes, so as to turn them with their heads towards the stockyard. Of course, he had to bring up the whole mob. My talents in the shouting line were not called out upon this occasion, for they all trotted into the stockyard of their own accord, and I had nothing to do but put up the slip-rail as fast as I could with only one available arm, for though it is better, I cannot use the other yet. When F— — came up we both went into the yard, and could soon make out the two horses which had their saddles on—that was the only way we could distinguish them in the dark. It was now nearly eleven o'clock, and though warm enough it was very cloudy, not a star to be seen. We fastened on the patched up bridles as well as we could by feeling, and mounted, and rode home, about three miles more, as fast as we could. When we entered the flat near our own house, we heard loud and prolonged "coo-ees" from all sides. The servants had made up their minds that some terrible misfortune had happened to us, and were setting out to look for us, "coo-eeing" as they came along. F— — pointed out to me, with a sort of "I-told-you-so" air, that there was no light in the drawing-room—so it was evident our friends had not arrived; and when we dismounted I found, to my great joy, that the house was empty. All our fatigue was forgotten in thankfulness that the poor travellers had not been exposed to such a cold, comfortless reception as would have awaited them if they had made their journey that day. I must tell you, they arrived quite safely the next evening, but very tired, especially the poor children; however, everything was ready, and the little boys were particularly pleased with their box beds, greatly preferring the difficulties of getting in and out of them to their own pretty little cribs at home. Such are boys all over the world!

Next month we leave this for ever, and go down to Christchurch to make our final arrangements for the long voyage of a hundred days before us. As the time draws near I realize how strong is the tie which has grown, even in these few short years, around my heart, connecting it with this lovely land, and the kind friends I have found in it. F — — feels the parting more deeply than I do, if possible, though for different reasons; he has lived so long among these beautiful hills, and is so accustomed to have before his eyes their grand outlines. He was telling me this the other day, and has put the same feelings into the following verses, which I now send you.

> A farewell. The seamen shout once and together, The anchor breaks up from the ground, And the ship's head swings to the weather, To the wind and the sea swings round; With a clamour the great sail steadies, In extreme of a storm scarce furled; Already a short wake eddies, And a furrow is cleft and curled To the right and left. Float out from the harbour and highland That hides all the region I know, Let me look a last time on the island Well seen from the sea to the snow. The lines of the ranges I follow, I travel the hills with my eyes, For I know where they make a deep hollow, A valley of grass and the rise Of streams clearer than glass. That haunt is too far for me wingless, And the hills of it sink out of sight, Yet my thought were but broken and stringless, And the daylight of song were but night. If I could not at will a winged dream let Lift me and take me and set Me again by the trees and the streamlet; These leagues make a wide water, yet The whole world shall not hide. Now my days leave the soft silent byway, And clothed in a various sort, In iron or gold, on life's highway New feet shall succeed, or stop short Shod hard these maybe, or made splendid, Fair and many, or evil and few, But the going of bare feet has ended, Of naked feet set in the new Meadow grass sweet and wet. I will long for the ways of soft walking, Grown tired of the dust and the

glare, And mute in the midst of much talking Will pine for the silences rare; Streets of peril and speech full of malice Will recall me the pastures and peace Which gardened and guarded those valleys With grasses as high as the knees, Calm as high as the sky: While the island secure in my spirit At ease on its own ocean rides, And Memory, a ship sailing near it, Shall float in with favouring tides, Shall enter the harbours and land me To visit the gorges and heights Whose aspects seemed once to command me, As queens by their charms command knights To achievements of arms. And as knights have caught sight of queens' faces Through the dust of the lists and the din, So, remembering these holiest places In the days when I lose or I win, I will yearn to them, all being over, Triumphant or trampled beneath, To this beautiful isle like a lover, To her evergreen brakes for a wreath, For a tear to her lakes. The last of her now is a brightening Far fire in the forested hills, The breeze as the night nears is heightening, The cordage draws tighter and thrills, Like a horse that is spurred by the rider The great vessel quivers and quails, And passes the billows beside her, The fair wind is strong in her sails, She is lifted along. THE END.

www.ingramcontent.com/pod-product-compliance
Lightning Source LLC
Chambersburg PA
CBHW070245230526
45470CB00002B/486